Work on the West Side

Work on the West Side

*Urban Neighborhoods and
the Cultural Exclusion of Youths*

Harald Bauder

LEXINGTON BOOKS
Lanham • Boulder • New York • Oxford

LEXINGTON BOOKS

Published in the United States of America
by Lexington Books
An Imprint of the Rowman & Littlefield Pu lishing Group
4720 Boston Way, Lanham, Maryland 20706

12 Hid's Copse Road
Cumnor Hill, Oxford OX2 9JJ, England

Copyright © 2002 by Lexington Books

All rights reserved. No part of this publication may be reproduced, stored in a retrieval system, or transmitted in any form or by any means, electronic, mechanical, photocopying, recording, or otherwise, without the prior permission of the publisher.

British Library Cataloguing in Publication Information Available

Library of Congress Cataloging-in-Publication Data

Bauder, Harald, 1969–
   Work on the west side : urban neighborhoods and the cultural exclusion of youths / Harald Bauder
      p.   cm.
Includes bibliographical references and index.
    ISBN 0-7391-0413-6 (cloth : alk. paper)
   1. Hispanic American youth—Employment—Texas—San Antonio. 2. Marginality, Social—Texas—San Antonio. 3. Inner cities—Texas—San Antonio. I. Title.
HD6275.S26 B38 2002
331.3'46368073'0976351—dc21

2002006016

Printed in the United States of America

∞™ The paper used in this publication meets the minimum requirements of American National Standard for Information Sciences—Permanence of Paper for Printed Library Materials, ANSI/NISO Z39.48–1992.

For my parents

Contents

Acknowledgments	ix
Introduction	1
1. Theorizing Work and Neighborhood	9
2. Mapping the Labor Market	25
3. Representing Neighborhoods: Views from the Inside and the Outside	41
4. "You're Good with Your Hands, Why Don't You Become an Auto Mechanic?"	55
5. Between Scales: Agency and Ideology	73
Conclusion	85
Appendix	91
Bibliography	101
Index	113
About the Author	117

Acknowledgments

Work on the West Side is the product of a long, laborious but rewarding process that involved many people. Without the support of these people I would have never been able to complete the research for this book, let alone write it. Most of all I thank Karen Uchic for her support throughout the writing and the production of this book. Her support ranged from encouragement to pursue a book project, to critically assessing the intellectual material of the book, to correcting my English.

Work on the West Side has its origins in my dissertation, which I completed at Wilfrid Laurier University. Although I took several years to revise and rethink the material, the book reflects the advice and guidance that I received from my dissertation committee. Bob Sharpe, the chair of the committee, has been a loyal supporter of this project from its very beginning. Sue Ruddick has been a greater influence on my thinking than she probably realizes. Tod Rutherford offered valuable feedback and pointed me toward important literature. George Galster identified critical gaps in my research but also provided me with the material to bridge them. Valerie Preston's constructive critique helped me move beyond conventional perspectives.

My colleagues in the Geography Department at the University of Guelph have been an extremely supportive group while I revised the manuscript. Richard Kuhn and Alun Joseph deserve special mention. During my two years as a postdoctoral fellow at the University of British Columbia, it was a true pleasure to work with David Ley; his advice and support provided me with the motivation to turn this project into a book. At Wayne State I received support from Robin Boyle, Eugene Perle, Gary Sands, Robert Sinclair, and Bryan Thompson. At various stages in the research process, I had insightful discussions with Miguel de Oliver and David Wilson. Comments on individual chapters were provided by Stuart Aitken, Robert Beauregard, Kevin Dunn, Derek Gregory, Doreen Mattingly, and Eugene Perle. William E. Doolittle made the facilities in the Geography Department of the University of Texas at Austin available to me.

While I was in San Antonio for fieldwork, Patti Radle, Luz Agah, Maria Maldonado, and Juan Diaz, invited me to participate in biweekly youth group meetings at Inner City Development, Inc., and to become a part of their community for a few months. Daniel H. Saucedo, Mike Tapia, Roberto J. Ramirez, Sgt. Victor Balderas, Jesse R. Chagoya, Josie Delgado, Rachele M. Guerrero, Gilbert Candia, Rod Radle, Alan J. Moy, Dave Sugg, Walter N. Augue, Joan Drennan-Taylor, and John Morris provided valuable insights that influenced this research. The project would not have been possible without the young people and administrators of local community organizations, who volunteered as interview respondents and informants. I promised to keep their personal information confidential. The two neighborhoods, which I call the Lanier area and Palm Heights, welcomed me as a guest. I admire the openness and hospitality of the residents of these two communities.

At Lexington Books, Jason Hallman oversaw the publication process. I also thank Martin Hayward and Brian Selzer. Marie Puddister at the Geography Department at Guelph put the finishing touches on the maps. I obtained additional cartographic support from Pamela Schaus at Wilfrid Laurier University. Kurt Metzger and Ruth Waite from the Michigan Metropolitan Information Center provided technical assistance.

The Association of American Geographers' Otis Starkey Fund Ph.D. Dissertation Grant, Wilfrid Laurier's Dean of Graduate Studies Ph.D. Dissertation Award, and the Externship for Dissertation Research from the Wilfrid Laurier University Graduate Students' Association financially supported the research for this book. I began preparing the manuscript while I was a Social Sciences and Humanities Research Council of Canada (SSHRC) postdoctoral fellow. I am indebted to the Geography Department at the University of Guelph, the Geography Department at the University of British Columbia, and the Department of Geography and Urban Planning at Wayne State University for offering their resources.

I thank several journals for granting me permission to revise the following previously published material: "Youth Labor Market Marginalization in San Antonio, Texas" (with Bob Sharpe), *The Professional Geographer* vol. 52, no. 3 (2000); "Agency, Structure, Scale: Representations of Inner-City Youth Identities," *Tijdschift voor Economische en Sociale Geografie* vol. 92, no. 3 (2001); "'You're Good with Your Hands, Why Don't You Become an Auto Mechanic': Neighborhood Context, Institutions and Career Development," *International Journal of*

Urban and Regional Research vol. 10, no. 1 (2001); "Culture in the Labor Market: Segmentation Theory and Perspectives of Place," *Progress in Human Geography* vol. 25, no. 1 (2001); "Work, Young People and Neighbourhood Representations," *Social and Cultural Geography* vol. 2, no. 4 (2001); "Neighborhood Effects and Cultural Exclusion," *Urban Studies*, vol. 39, no. 1 (2002).

Although many people have left an imprint on this book the viewpoints presented here are my own and do not necessarily reflect the opinions of others.

A Note about the Cover:

The mural *Familia y Cultura es Vida* is located on San Antonio's West Side and was blessed and dedicated on May 6, 1995. The San Anto Cultural Arts (www.sananto.org) writes:

> Familia is the best example of direct documentation, through muralism, of barrio culture as it existed at that particular place and time. The mural is located on a building that, up until 1999, was home to Westside resident and neighborhood mechanic/accordionist, Paul Morada. On weekend nights, Paul and friends, collectively known as Los Guadalupanos (in honor of their dedication to the Virgen De Guadalupe and the street where their music and art was created, Guadalupe St.), would gather on the front porch of Paul's home/mechanic shop and play Conjunto music till the wee hours of the night. Mr. Morada would go from rebuilding a carbeurator to reconstructing "Tu Traicion" by Cornelio Reyna in just a few hours. The mural was inspired by this ritual and depicts one such scene as visualized through the eyes of the artists: twelve-year-old Debbie Esparza and co-founder Juan Ramos.

I selected this mural as the cover because it illustrates how cultural and labor market identities are intertwined. The scene of the auto mechanic (left side of the mural) resonates with the representation of local workers as discussed in Chapter 4. I thank the San Anto Cultural Arts, Inc., for the use of the mural *Familia y Cultura es Vida*. All net royalties from *Work on the West Side* will be donated to San Anto Cultural Arts, which involves the local community in a series of cultural arts projects.

Introduction

My first confrontation with the issue of inner-city marginality took place in the infamous Cass Corridor in Detroit, where I unwittingly moved in 1991 to assume an athletic scholarship at Wayne State University. Spending my entire life until then in Germany, I was relatively untainted by the discourse of race and the inner city that circulated in the United States. I did not see the pathology that supposedly haunted my neighbors as outlined in my courses (through books like William J. Wilson's [1987] *The Truly Disadvantaged*). Rather, to me as a newcomer, some of the behavior among my middle-class suburban classmates was just as strange as that of my inner-city friends. What I did notice, however, was the antagonism between the two groups that bubbled below the surface of friendly courtesies and superficial small talk. Observing the interaction between my fellow students in the classroom, in my student apartment building, or in off-campus taverns, it occurred to me that the different groups that were brought together on Wayne State's urban campus knew relatively little about each other, and that an inner-city neighborhood, like the one in which I lived, can be a terribly frightening place to someone who grew up in a 'middle-class' suburb.

In this book, I explore issues of cultural representation of neighborhood communities. The social and economic exclusion of minorities in U.S. inner cities has been the focus of public debate for decades (Harris and Curtis 1998). But despite extended research efforts by urban anthropologists, economists, geographers, sociologists, and planners we still know very little about the role of cultural representation of inner-city neighborhoods in the production of marginality. In addition, most of the things that we do know about inner-city marginality involve African Americans in cities like Detroit, New York, or Chicago, or Latino immigrants in Los Angeles or Miami. A book on representations of urban Latino communities in San Antonio, Texas, is long overdue. I argue that cultural judgments play a crucial role in the ongoing marginalization of inner-city communities. My aim is to extend existing perspectives of inner-city exclusion, which emphasize economic and social processes, to the cultural realm.

Culture is a difficult idea to wrestle with. According to Raymond Williams (1983, 87), *culture* is "one of the two or three most complicated words in the English language." The term is inconsistently applied across disciplines and is continuously rethought and reinvented. In the context of this book I treat culture as a system of signification that produces meaning through the interpretation of behavioral patterns and material attributes. I reject essentialist perspectives that define culture as an absolute and fixed set of "customary beliefs, social forms, and material traits" (*Webster's Dictionary*). "There's no such thing as culture," Mitchell (1995, 109) proclaims, but "'culture' is an ideology." Cultural differentiation orders people in a manner that relates to our everyday experiences of the social world. In the debate about inner cities, the conception of culture often reflects the particular worldviews of commentators, politicians, and social scientists.

To create and uphold a cultural ideology that devalues urban minority communities requires a great deal of human effort. But it would be naïve to suggest that a single author or a particular group of agents have plotted and designed this system of cultural hegemony. Rather, cultural difference is discursively scripted involving multiple authors and institutions. Cultural difference infiltrates common social practices and becomes a part of our "taken-for-granted world" (Ley 1977). The process of globalization, for instance, has produced the effect of "overvalorizing" the work performed by a professional and management elite, while "devalorizing" the labor of manual and low-skill workers. This effect has widened the gap between the urban middle class, which experiences an upgrading of their social status, and less privileged inner-city residents, who suffer from status decline (Ley 1996: Sassen 1994). Discursive processes have made transnational corporations, ordinary citizens, well-intending activists, and unassuming social scientists co-participants in the social and economic exclusion of inner-city communities.

Cultural representations of inner-city communities are not fixed but are continuously contested and redefined. These dynamics open the possibility for intervention and reshaping of current ideas of urban minority communities. This book does not outline a future vision of how these communities should be represented or how they should represent themselves. As an outsider,[1] I cannot speak in this regard for an inner-city Latino community in San Antonio. Rather, I attempt to juxtapose internal and external representations to dismantle some of the elements that constitute the discourse of inner cities. My aim is not to shape but merely to enable fresh imaginaries of the inner city and its residents.

The cultural exclusion of inner-city communities is intimately related to labor market processes. The concept of cultural capital has helped me grasp the linkage between cultural representation and the value of labor. Cultural capital expresses value in the form of style, taste, behavior, and the possession of objects (Bourdieu and Passeron 1977). It relies on shared cultural codes to interpret what is valuable and what is not. Objects, such as paintings, "do not impose the self-evidence of a universal, unanimously approved meaning" (Bourdieu 1984, 100). Rather, their exchange value is "only revealed in the world of social uses . . . and . . . the schemes of perception, apprehension and action" (Bourdieu 1984, 100). Similarly, workers can be evaluated for the exchange value of their labor based on the cultural markers they display. Physical appearance and behavior, for example, are markers that require shared meanings to be decoded and expressed as value in the market. Cultural codes, however, vary between communities and geographical contexts, assigning different values and meanings to the same cultural symbols. Urban labor markets span across the boundaries of neighborhood communities and bring conflicting coding schemes in contact with each other, with the effect of devaluating the cultural capital of disenfranchised communities. In addition, neighborhood of residence itself functions as a symbolic marker that signifies a worker's value in the labor market.

But there is more to cultural representation than expressing monetary exchange value in the labor market. Work identities are an important part of a person's cultural identity (Fernández Kelly 1994; Harvey 1990). Being an auto mechanic or a doctor defines not only wage ranges, it also shapes a person's self-identity and membership in a community. Especially among young people, entrance into the labor market often becomes synonymous with "crafting an identity" (Newman 1999, 88). Paul E. Willis (1977), in *Learning to Labor*, illustrates this point. He describes a group of English working-class lads who affirm their masculinity by rejecting "school culture" and entering into manual labor occupations. Their labor market roles make them legitimate members of a particular peer group and the general working class. What Willis does not consider, however, is how spatial context and the neighborhood community can frame these processes of identity formation.

Social life is spatially fragmented and locally situated. "There are no aspatial social processes. Even in the realm of pure abstraction, ideology, and representation, there is a pervasive and pertinent, if often hidden, spatial dimension" (Soja 1996, 46).[2] At the same time, the way we think about space is itself an ongoing cultural process (Gregory 1994). In a famous quote, Lefebvre (1991, 26) states: "(Social) space is a (social) product." The inscription of marginality onto inner-city space, through the continuous extraction of images relating to crime, drug addiction, and gang warfare, has disempowered these communities and their residents (Wilson 2001). The inner-city neighborhood has become a symbol of social dysfunction, and the behavior, life priorities, and work aspirations of young people in these neighborhoods serve as the markers of pathology. Spatial processes of identity formation and representation are also at the heart of processes of labor market exclusion of inner-city youths.

The spatial perspective of cultural processes assumed in this book aims to expand the existing debate on inner-city marginality. Much of this debate has relied on essentialist conceptions of culture as a euphemism for ethnicity, race, and class, and it typically distinguishes between the cultural "mainstream" and "dysfunctional" inner-city neighborhoods.[3] In the 1960s, the culture-of-poverty concept asserted that a subculture of "a strong feeling of fatalism, helplessness, dependence and inferiority, . . . weak ego structure, orality and confusion of sexual identification . . . a strong present-time orientation with relatively little disposition to defer gratification and plan for the future" (Lewis 1966, 23) in combination with out-of-wedlock childbirth (Moynihan 1965) caused poverty among urban minorities. The proposed solution was the gradual assimilation of the pathological community to the norms of an imaginary mainstream. Conservatives even suggested that welfare-related services reinforce poverty culture and therefore argued for the withdrawal of services altogether (Murray 1984). Although the underclass concept attributes the initial trigger of urban poverty to structural changes in the urban economy, it continues to promote the idea that a subsequent cycle of social pathology drives the community into economic desolation (Wilson 1987).[4]

But the 'culture' of inner-city minority communities should never be situated outside its wider social context. Steinberg (1981, preface) stresses that the cultural identity of one group is necessarily defined relative to another group:

> That ethnic groups have unique cultural character can hardly be denied. The problem, however, is that culture does not exist in a vacuum; nor is it fixed or unchanging. On the contrary, culture is in constant flux and is integrally a part of a larger social process. The mandate for social inquiry, therefore, is that ethnic patterns should not be taken at face value,

but must be related to the larger social matrix in which they are embedded.

Therefore, one cannot uncritically accept the notions of the cultural "mainstream" and "dysfunction" as absolute categories in the debate of economic marginality of inner cities. If we did this, we would derive conclusions from the false assumption that inner-city communities are closed systems detached from their wider sociopolitical circumstances. In this case, blame could easily be affixed to the victims themselves, legitimizing the ongoing processes of the social exclusion of inner-city minorities (Bourgois 1995; Fainstein 1993; Gans 1990). Academic debate of inner-city marginality must therefore open itself to a healthy dose of self-criticism and examine the cultural relationship between inner-city communities and wider society. This self-criticism must include our perspective of the labor market. We must recognize that the value of workers cannot be measured by a universal yardstick, but that ideas of achievement and status differ between social contexts. It is improper to apply the same measures of success and failure to middle-class suburban areas and to minority inner-city communities.

Young people are a major object in the exclusionary discourse of the inner city. They are often labeled underachievers, street kids, and predatory gang members. But young people are also a particularly vulnerable population group. Youth is a period when adult self-identities are negotiated, and young people make important and long-lasting life choices regarding education, skill attainment, and entrance—or non-entrance—into the labor market (Bowlby et al 1998; Irwin 1995). Yet, agency to shape and challenge these identities is severely constrained during this life stage (Valentine et al 1998). Ruddick (1996, 3) explains:

> [Adolescents] are denied agency in all but the most banal forms. . . . Adolescents, in the common view, are generally considered too old to be ascribed the power of "nature"—we do not look on their activities with awe and wonder, the way we do those of small children. Yet, adolescents are generally considered too young to be reasoned actors in the sense one might consider adults. When it comes to any form of sustained and serious agency, adolescents are depicted as awkward, simple-minded—"stupid and contagious." They live in a state where agency is continually denied them.

Kett (1977) suggests that the category "adolescence" was created in the first place to restrict agency and to control the reproduction of class structure. With their agency constrained, youths are especially exposed to the consequences of cultural representations of their residential space. The following chapters examine the role of neighborhood representations in the formation of work identities of young inner-city Latinos and Latinas in San Antonio.

The notions of work and labor market identity, in the context of this book, refer not only to the occupational roles and aspirations of young people, but also to the perception of what constitutes desirable, sufficient, and inadequate employment. Chapter 1 develops the two theoretical arguments that relate issues of spatial representation to work and labor market identities of young people. The first argument presents an interpretation of labor market segmentation theory that links the problem of segmentation to spatial representations of the inner city. The second argument examines how the idea of neighborhood effects contributes to the making of urban marginality, and how it legitimizes the labor market exclusion of young people who live in inner cities. These two arguments frame the empirical study in the remaining chapters.

Chapter 2 introduces the urban and demographic context of San Antonio and presents a statistical analysis that measures the degree of youth labor market marginalization by neighborhoods. This analysis also assesses the quantifiable linkages between the social circumstances of the neighborhood and youth marginality. Based on this analysis, two neighborhoods, the Lanier area and Palm Heights, are selected for an in-depth ethnographic study to investigate causal processes of cultural representation that enable the interpretation of the emergent quantitative patterns.

Chapter 3 identifies the work identities of young people in the Lanier area and Palm Heights. Internal and external work identities are juxtaposed, revealing important differences between the two neighborhoods. For instance, "flipping hamburgers" is not considered marginal work in the Lanier area, while in Palm Heights young people aspire to professional careers. Such differences render it inappropriate to apply a universal scale of labor market segmentation to the different cultural contexts.

Chapter 4 is concerned with community-based institutions and their role in influencing the work identities and labor market expectations of young people. To identify problem areas and design strategies of response, institutions interpret neighborhood context and deploy their own cultural judgments. One institutional administrator recalls a general sentiment against Lanier area youths: "You're good with your hands, why don't you become an auto mechanic." Such neighborhood-based interpretations of work roles contribute significantly to the labor market marginalization of young people.

The final empirical chapter focuses on the Lanier area to explore the issue of scale in the representation of inner-city dysfunction. I suggest that ideological perspectives of work and career occupy overlapping scales of place. This scale dependency enables the exclusion of young people based on their neighborhood of residence. While "jumping" from the neighborhood scale to a wider ideological scale of reference may lead to improved employment prospects for young people, it does not necessarily constitute an act of culture empowerment, but rather a concession to dominant cultural norms.

In this book, I take a political stand. My description of concrete social practices and behavior may isolate some agents and institutions within a sequence of causal events that lead to marginality. It is not my intention to blame particular agents or institutions. Rather, my aim is to encourage constructive communication between communities, institutions, and scholars, and invite them to participate in the project of re-scripting the discursive politics of place and scale that have dispossessed minority communities in America's inner cities.

Notes

1. I am a 'white' male in my early thirties who grew up in a suburb of a mid-size city in southern Germany. Since leaving my hometown, I have lived in the industrial Ruhrgebiet in Germany, Detroit, San Antonio, Vancouver, and Southern Ontario, where I currently reside.

2. Most geographers have grown suspicious of a universally applicable "grand theory." Their research stresses locally situated knowledges and interpret society from a concrete, place-based vantage point (Gregory 2001).

3. Kelly (1997, 15-42) illustrates that social scientists who claim to have discovered authentic African American culture have indeed created simplistic and misleading images of people and their communities.

4. Even in the current climate of multiculturalism, the recognition of ethnic communities is hinged on profits in the fashion, entertainment, and tourist industries, which sell images of exoticism (e.g., de Oliver 2001).

Chapter 1

Theorizing Work and Neighborhood

Labor Segmentation and Cultural Representation

In the world of neoclassical economics, the labor market is the perfect matching mechanism of workers and jobs. Workers acquire education and job skills that suit the needs of employers, and employers offer the wages and benefits that attract an appropriate workforce. The neoclassical perspective of the labor market is democratic because the rules of the market are universal, spanning across occupations and social categories. In the world of political economy, on the other hand, the labor market is not as democratic as neoclassical theory suggests. Workers and jobs are divided into so-called labor market segments (Averitt 1968; Gordon et al 1982; Reich et al 1973; Wilkinson 1981). The boundaries between segments are rigid, ensuring that workers and jobs are matched within segments and that "the rules governing the behavior of labor market actors differ from one segment of the labor market to the other" (Peck 1996, 46). To these two perspectives, a cultural perspective of the labor market must be added to understand how workers are segmented according to the neighborhoods in which they live.

Before I return to this cultural perspective, I want to discuss the spatiality of labor segmentation. Labor segmentation is the product of both demand and supply side processes (Fevre 1992, 1-22). A demand-side perspective—focusing on the characteristics of jobs, such as skill and educational requirements, employment stability, and wages—suggests that the exclusion of inner-city minorities from the primary labor market segment is an effect of the spatial dynamics of industrial restructuring (Kasarda 1990; Skinner 1995). The spatial mismatch hypothesis, for instance, argues that manufacturing jobs and employment with decent wages have moved to the suburbs and are no longer available in inner-city neighborhoods. Inner-city residents are therefore pushed into the secondary labor market or into unemployment.[1]

A supply-driven perspective recognizes that workers are social actors as much as they are labor. This perspective suggests that labor supply characteristics, such as education, skills, and occupational preferences of workers, reflect wider social processes rather than pure market forces. Thus, social divisions are constructed outside of the labor market, but then translate into labor segmentation. For example, the social roles of the female child rearer and the male breadwinner create gendered identities that channel women into secretarial or nursing occupations and men into manufacturing or construction sectors (Kenrick 1981; Kessler-Harris 1982; Willis 1977). Similar social processes segment workers by their class and ethnicity (Morales and Bonilla 1993; Offe and Hinrichs 1985; Segura 1995). Supply-side approaches to labor market segmentation emphasize: "One reason secondary work exists is in the *prior* existence of a group of workers who can be exploited in this way" (Peck 1996, 69; original emphasis).

The linkage between supply-driven social differentiation and labor segmentation is not a one-directional cause-effect relationship. Rather, social inequality is the result of a set of interlocking policies and practices circulating through a variety of social institutions, including the labor market (Hanson and Pratt 1995). Max Weber (1968, 928) realized that work and social meaning are mutually dependent, and that "'class situation' is ultimately 'market situation.'" Social hierarchy is produced and maintained by controlling the processes that allocate workers to jobs, but occupational status also shapes workers' social position (Parkin 1974). These processes have a spatial dimension. Spatial mismatch, for example, begs the question why the residences of some groups are spatially removed from employment opportunities in the first place. The labor market contributes to residential inequalities. Families with low incomes, for instance, simply cannot afford to live in upscale neighborhoods, and often have no choice but to live in class-segregated neighborhoods.

A cultural perspective of the segmentation of labor allows for agency in the creation of labor market identities of workers; an idea that still remains underdeveloped in contemporary labor market segmentation theory (Marsden 1986). Cultural identity can be self-ascribed or imposed by others. The latter represents culture as an external construct of domination and "otherness" (Omi and Winant 1986). The former refers to culture as a lived experience that is intrinsic to a community. Internal and external cultural representations are conceptually distinct processes that create different cultural boundaries and affect labor market segmentation in particular ways.

The everyday experience of social life creates cultural capital that articulates internal group identities (Bourdieu 1984). This cultural capital "constitutes a repertory of symbols (that) affect the relationship between individuals, social networks and economic structures, including labor markets" (Fernández Kelly 1994, 100). A system of symbolic markers and cultural codes, associated with behavior, bodily signifiers, and material traits, expresses internal group identity and enables individuals to maneuver within the cultural context of the group. Similarly, this system of symbolic markers and cultural codes enables the internal functioning labor market segments. For example, the working class identifies itself not only through affiliation with blue-collar work but also through distinct working-class behavior and attitudes (Willis 1977). On the upper end of the labor market, in corporate banking, proper behavior and physical appearance are also required for the inclusion into this elite occupational category (McDowell 1997). In the service industry, restaurant employees often work in the "front" as waiters, bartenders, or hosts, or in the "back" as cooks and kitchen aids, depending on whether or not their repertoire of symbols matches the cultural requirements of the work task (Crang 1994; Zukin 1995, 153-185). Cultural capital facilitates the segmentation of labor.

Cultural capital is linked to the social, political, and historical contexts of place. *Structures of feeling*, a concept developed by Raymond Williams (1958), has been interpreted as a cultural quality of place that mediates local identity (Jackson 1991; Longhurst 1991). Williams (1989, 242) states that "place has been shown to be a crucial element in the bonding process." Neighborhood-based structures of feeling ground self-identity and produce symbolic meanings that situate workers in a segmented labor market.[2] Work identities are "broad cultural and more specific *local* constructions of 'who should do what and where'" (Cope 1998, 138; my emphasis). Accordingly, perceptions of what constitutes good and bad jobs differ between neighborhoods (Wial 1991). Neighborhood context also shapes the negotiation between motherhood and paid employment (Holloway 1999). In American inner cities, the experience and interpretation of neighborhood-particular symbols and meanings frame the labor market identities of young people. For example, having children as a young adult, discontinuing education, and entering the labor force at an early age (all supposedly dysfunctional underclass traits) can signify adulthood, independence, and achievement in these local contexts (Fernández Kelly 1994).

Young people are initiated into their social status categories and socialized into labor market identities through educational institutions (Kett 1977). The school

> is where working-class themes are mediated to individuals and groups in their own determinate context and where working-class kids creatively develop, transform and finally reproduce aspects of the larger culture in their own praxis in such a way as to finally direct them to certain kinds of work. (Willis 1977, 2)

Schools are local institutions: they tend to draw students from particular residential areas and help define neighborhood identities. Other local institutions that mediate cultural meanings of work include community centers, churches, and government authorities.

In addition to local institutional context, cultural meanings of work are inevitably intertwined with wider social, political, and historical circumstances of residence. "Residential segregation . . . creates varying social milieux which foster distinctive working-class subcultures, structured along the lines of ethnicity, stage in the life cycle and levels of skill. Educational aspirations and resources vary across these milieus, and this has the effect of reproducing skills across generations" (Pratt 1989, 101). Neighborhood-specific work identities and cultural experiences introduce a supply-side argument to labor segmentation theory that stresses the importance of residential inequalities and local particularity.

Another force that influences labor segmentation relates to external processes of cultural differentiation. For example, external labels permit U.S. employers to view young Latinos as cheap and "disposable" labor (Goldberg 1997). Likewise, Latina women have acquired stereotypical identities as housekeepers, cleaners, and sweatshop workers (Cross et al 1990; Menchacca 1995). Just like ethnicity situates workers into labor market segments, so can the cultural status of a neighborhood indicate the suitability of residents to perform certain kinds of work.

Discourses and "common sense" social practices (Smith 1989) of exclusion and inclusion produce external images of neighborhoods that project categorical cultural qualities onto residents.[3] The popular depiction of Vancouver's Chinatown exemplifies this process of external neighborhood representation:

> As an idea, one that relied on a range of cultural assumptions held by Europeans about the Chinese as a type . . . it was an evaluative term. . . . Regardless of how Chinatown's residents defined themselves and each other—whether by class, gender, ethnicity, region of origin, surname,

generation, dialect, place of birth, and so on—the settlements were perceived by Europeans through lenses of their own tinting. Without needing the recognition of residents, Chinatown's representers constructed in their own minds a boundary between "their" territory and "our" territory. (Anderson 1991, 30)

In a similar vein, perceptions of criminal threat are racially and ethnically coded and shaped by the number of Latinos and African-Americans living in an urban neighborhood (Chiricos et al 2001).

If such ideas of neighborhoods express external labor market identities, and if employers, educators, and labor market regulators rely on these identities to categorize workers into labor market segments, then neighborhood representation becomes an important factor in the segmentation of labor. The derogatory labels of "ghetto" and "barrio" impose a cultural identity on neighborhoods that suggest low skill level, poor work ethic, and occupational incompetence guiding employers' and educators' expectations of residents (Holzer 1996; Kirschenman and Neckerman 1991; Lee and Wrench 1987; Waldinger 1997). Some employers blatantly discriminate against inner-city residents (Turner 1997; Wilson 1996). For example, the personnel manager of a temp agency in suburban Detroit recently filed a complaint at the regional office of the Equal Employment Opportunity Commission because a client demanded "no Detroit residents" in its recruitment profile. It is illegal to screen temp workers by race, gender, religion, and disability, but the agency allowed clients to filter out Detroit residents from the pool of eligible workers (Hopgood 2000).

While neighborhood stereotypes often exclude local residents from the labor market, they sometimes justify the recruitment of workers into certain occupations. Some employers search in immigrant reception areas for a vulnerable workforce that is willing to endure substandard working conditions. Others look in suburban areas for an educated labor force of middle-class housewives who can be exploited as part-time workers (England 1993). That employers consider neighborhood of residence as a recruitment factor confirms the importance of cultural representation of place of residence in the segmentation of labor.

One cannot easily separate residential segregation, cultural representation of neighborhoods, and labor market segmentation into cause and effect. Just as cultural representation of place influences the segmentation of labor, so does a spatially segregated labor force give rise to the construction of neighborhood-based stereotypes. A large public housing project, for instance, pools workers with low skills and few educational credentials. Once such a group of people is residentially isolated, their neighborhood acquires a stigma that further disadvantages them in the labor market. Cultural differentiation, residential segregation, and economic segmentation are interlocking processes. Located at the center of these processes, however, is the neighborhood, which assumes a constitutive role in the segmentation process.

Cultural differentiation and neighborhood representations are, of course, social processes that involve human agents and decision makers. Politicians, planners, community workers, and academics are not innocent bystanders to the neighborhood-based segmentation of labor. Rather, they participate actively in the creation of external cultural identities of inner-city neighborhoods. Academic research in the social sciences, for example, rarely stops at stating simple objective facts, but it interprets these facts in light of the ideological worldviews of the researcher and the academic community. In the context of inner-city research, academic discourse has engaged in the production of cultural labels that depict inner-city communities as socially pathological space. This discourse legitimizes the labor market exclusion of inner-city residents and lays the blame of marginality to the inner-city community itself. In the next section I examine how academic debate has played an important role in the making of dysfunctional inner-city space.

The Making of Inner-City Dysfunction

An influential study of the Gautreaux Assisted Housing Program in Chicago monitored the participants of a housing voucher program that gave poor families with children a choice to move into suburban residences or to remain in minority, inner-city neighborhoods. The study revealed that the children of suburban movers went on to perform better in school and in the labor market than their inner-city counterparts (Rosenbaum 1991, 1995). The author of this study concluded that middle-class suburbs provide a cultural environment in which young people learn important social and behavioral skills not available in poor inner-city neighborhoods.

These so-called neighborhood effects (Galster and Killen 1995; Gebhart 1997; Jencks and Mayer 1990) are believed to operate through three mechanisms. The first mechanism, peer group influences, proposes that local peer networks "infect" youngsters with pathological behavior. Researchers have used the metaphor of the "epidemic" to suggest that social dysfunction spreads through peer networks like a deadly disease (Crane 1991b). The second mechanism, collective socialization, suggests that local adults are role models to young people. In "concentrated poverty" areas, a large number of local adults possess traits that are presumed to be pathological, such as out-of-wedlock births, crime, poverty, and male nonemployment. These traits are passed on to young people resulting in "problematic behavior," including teenage childbearing, school dropout, and labor market failure (Galster and Mikelsons 1995; O'Regan and Quigley 1996). Mainstream neighborhoods, on the other hand, are defined in terms of "residential stability" (Sampson et al 1997) or "the percentage of workers in the neighborhood who held professional or managerial jobs" (Crane 1991a, 303). Such neighborhoods are believed to result in unproblematic behavior among young people. The third mechanism focuses on a dilapidated physical infrastructure that destabilizes communities[4] and on local institutions, such as schools, that commonly fail to provide adequate services in poverty-stricken areas (Kozol 1991; Waggoner 1991).

My critique targets the first two explanations of neighborhood effects: peer group influences and collective socialization effects. These two explanations essentialize culture. They imply that the social context of poor inner-city neighborhoods instills dysfunctional family norms, disinterest in education, and poor work ethic into young people, and thereby triggers a cycle of social pathology and dysfunction. The idea of neighborhood effects offers a seemingly nonracial and class-free interpretation of neighborhood marginalization, but instead focuses attention on the cultural attributes of communities. Many urban researchers embrace this explanation of inner-city marginality. They usually do not acknowledge that this explanation imposes labels of cultural dysfunction on neighborhoods and blames communities for their own marginality.

The idea of neighborhood effects is part of a wider urban underclass discourse that seeks to explain poverty among urban minorities through a pathological conception of inner-city culture.[5] This underclass discourse creates the impression that dysfunctional cultural norms, values, and behaviors are autonomous factors which produce social and economic marginality (Fainstein 1993; Gans 1990). The idea of neighborhood effects extends this logic and implies that cultural pathology circulates through the neighborhood-based community and ultimately causes the economic failure of the community's members. The appropriation of geographical place on the neighborhood scale makes neighborhood effects a particularly powerful idea within the underclass discourse. The neighborhood, not personal characteristics, signifies social pathology and dysfunction. People are marked not because of who they are or what they do, but because of where they live. This viewpoint is politically less confrontational because it does not blame ethnic and racial minorities but rather the anonymous context of the neighborhood.

Not only do neighborhood effect studies provide scientific legitimacy to processes of neighborhood-based exclusion, but they have shaped public policy and planning initiatives. For instance, the U.S. Department of Housing and Urban Development (HUD) supports initiatives for the residential dispersal of low-income families. One such initiative is the Moving to Opportunities (MTO) pilot project, which is designed to replicate the Gautreaux program in Baltimore, Boston, Chicago, Los Angeles, and New York City. MTO has been in operation since March 17, 1994 (Rosenbaum 1995; U.S. Department of Housing and Urban Development 2001). HUD also awards Section 8 housing vouchers to 1.3 to 1.4 million low-income families annually. Under Section 8, participating families are responsible for finding a suitable housing unit and HUD pays a housing subsidy to the landlord on behalf of the family. This subsidy enables low-income families to move to middle-class and suburban areas (U.S. Department of Housing and Urban Development 2001). Regional HUD initiatives and local housing authorities offer similar programs. Implicit in these efforts is the message that moving poor families into middle-class areas eliminates negative neighborhood effects (Briggs et al 1999; Nenno 1998; Rosenbaum and Harris 2001; Varady and Walker 2000).

Other strategies include mixed-income housing strategies, which seek to transform the cultural environment of neighborhoods and thereby offset negative neighborhood effects (Brooks-Gunn et al 1997; *Cityscape* 1997). HUD's HOPE VI program, for instance, seeks to create "economically integrated" inner-city communities. Urban enrichment schemes have a similar objective. HUD awards so-called Community Development Block Grants to metropolitan regions and urban counties to revitalize neighborhoods. Among the stated eligible activities of this program is to establish "neighborhood watch programs, providing extra police patrols . . . and clearing abandoned buildings used for illegal activities" (U.S. Department of Housing and Urban Development 2001). While the formal objective of these initiatives is to provide direct assistance to poor communities and to promote social and economic integration, an implicit aim is to discourage the formation of internal cultural identities within neighborhoods that are pathological and associated with educational and labor market failure.

I challenge, in particular, the use of the neighborhood effects model to explain school dropout, welfare dependency, and labor market marginality. Although statistical evidence clearly demonstrates that neighborhood characteristics are correlated with individual outcomes, the causality of neighborhood effects remains ambiguous. We do not know how neighborhood effects really work. Statistical analysis reveals correlations and describes quantitative patterns, but it does not explain *why* neighborhood circumstances and individual outcomes are correlated. Two of the most prominent supporters of neighborhood effects remark:

> Almost all of [the evidence on neighborhood effects] relies on a "black box" model of neighborhood and school effects that makes no assumption about how social composition influences individual behavior. Models of this kind try to answer the question, How much would an individual's behavior change if he or she moved from a low-[socioeconomic status] to a high-[socioeconomic status] neighborhood or school? They do not purport to explain *why* moving has an effect. (Jencks and Mayer 1990, 115; original emphasis)

Galster and Mikelsons (1995, 74-75) agree that a "fundamental challenge . . . is to distinguish measures of causal input and behavioral outputs. . . . This problem is particularly acute when one tries to understand the relationship between *aggregations* of individual behaviors and neighborhood-level socioeconomic conditions that affect the behaviors of *individuals*" (original emphasis). Statistical peer group effects, for example, tend to disappear when the statistical model is modified to account for the fact that these groups are selected by youths themselves and the residential choices of their parents (Evans et al 1993). "Individuals select neighborhoods, and neighborhoods their residents, by virtue of the goods and services they offer and the characteristics of their residents" (Furstenberg et al 1999, 19-20). In addition, presumed dysfunctional behavior exists in middle-class neighborhoods as well, but without the devastating effects it supposedly has in the so-called concentrated poverty neighborhoods (Pattillo-McCoy 1999). The implied causality of neighborhood effects is a common fallacy of research that examines inner-city marginality.

Without understanding the causal mechanisms of how neighborhood effects actually work, statistical results are ambiguous. Crane (1991a, 303), who studied the effect of neighbors' occupational characteristics on youths, admits: "Just why the percentage of workers who held high-status jobs had the strongest effect is not entirely clear." Rosenbaum (1991, 1205), who measured the educational and labor market capabilities of city-to-suburb movers, notes: "It is hard to know exactly what features of the suburbs allow this to happen." Indeed, the logic of the neighborhood-effects argument is circular: if the source of marginality is also the outcome—i.e., if socially marginalized communities also host socially marginalized people—then neighborhood effects cannot be statistically verified. In light of this ambiguity, Turner and Gould (1997, 64) lament the lack of ethnographic research to "identify and test the causal mechanisms that link neighborhood conditions to individual outcomes."

Through sanctioning the idea of neighborhood effects, social scientists actively participate in an ideological discourse that valorizes professional work aspirations, the nuclear family, and sexual abstinence during adolescence, while denouncing behavior such as school dropout, welfare dependency, and teenage and out-of-wedlock childbirth as inherently pathological.[6] But these "underclass" traits are not automatically harmful; they do not *necessarily* cause economic marginality (Gans 1990). Similarly, the working poor often embrace the same family values and attitudes toward employment as more affluent groups, but they nevertheless suffer from economic hardship (Newman 1999).

Underprivileged communities, of course, do not actively isolate themselves from society and economic opportunity. Rather, they are marginalized because of their cultural differences. This realization exposes the ideological underpinnings of the idea of neighborhood effects. Neighborhood effects research does not occur "in an epistemological, moral, or ideological vacuum. . . . On the contrary, public policy decisions should be normative, with ethics and democratic philosophy playing important roles in policy research" (Galster 1996, 2). Social scientists participate in deciding "what phenomena are defined 'social problems,' what their severity and consequences are, and what range of governmental responses is appropriate" (8). It is the specification of social problems and the identification of dysfunctional neighborhoods that open the idea of neighborhood effects to critique.

The idea of neighborhood effects is yet another episode in the ongoing discourse in American politics and urban research that affixes causality of labor market marginality to poor communities themselves. The culture-of-poverty concept of the 1960s (Lewis 1965, 1966; Moynihan 1965) and the more recent underclass debate (Wilson 1987) both imply that deficiencies procreate through the community and cause economic marginality. Some researchers even measure "concentrated poverty" by high school dropout rate, proportion of males with no stable employment, welfare dependency rate, and proportion of female-headed households (Hughes 1989; Kasarda 1993; Ricketts and Sawhill 1988), indicating that the sum of these behaviors defines an extreme form of poverty. This position, however, "confus[es] behavior with economic outcomes" (Jargowsky 1997, 24).

How exactly do neighborhood effects operate? Neighborhood effects are born out of a combination of local processes of identity formation and outside perceptions of the community. I call these influencing factors internal and external neighborhood representations. Regarding internal representations, residents commonly identify with the cultural environment of their neighborhoods. In fact, many low-income households who receive housing vouchers decide not to move into alienating upper-status areas, but prefer to stay in or close to their previous neighborhood (Pendall 2000; Varady and Walker 2000). Neighborhood context offers personal networks that are important "survival strategies" for obtaining housing, child care, and employment (Gilbert 1998). These strategies often compensate for the denied services and lack of labor market opportunities, and they enable communities to maintain their cultural identity (Dunn 1998; Peach 1996). Child rearing outside of the context of a nuclear family, for instance, may not be a terribly defiant practice in neighborhoods in which extended family and community support networks exist. This behavior only becomes "harmful" to labor market prospects if appropriate child care is unavailable. Likewise, low-wage employment need not necessarily infringe on a person's livelihood if tight community networks offer goods, services, and benefits outside the formal economy (Gibson-Graham 1996). A fallacy of the neighborhood effects idea is that it applies universal norms of child rearing, school performance, and labor market success to culturally distinct neighborhoods, although not all communities adhere to the same norms of family, education, and work.

Moving families out of inner-city neighborhoods may indeed change the educational and labor market outcomes for the children in these families, as predicted by the neighborhood effects model. However, the reason for improvement is not that the movers become better families and individuals, but that they assimilate to a dominant set of cultural norms and, therefore, suffer less from processes of cultural exclusion in the school system, government institutions, and the labor market.[7] Communities that adhere to alternative family norms, lifestyles, and perspectives of work are not given the same chances as communities that fit the mythical model of 'mainstream' American society.

Neighborhood effects also operate through external neighborhood representations, which often come in the form of cultural stereotyping. Media depictions of the inner city as pathological space have shaped public perceptions of inner-city neighborhoods as inherently prone to economic failure and labor market marginality (Wilson 2001). This perception reinforces processes of cultural exclusion of neighborhoods that are marked dysfunctional. The academic literature legitimizes these processes by offering a "straight-forward" explanation of neighborhood effects that is intuitively appealing because it claims to be objective and value-free. But without grasping how exactly neighborhood effects operate, external neighborhood representations augment the problem of cultural exclusion in the labor market.

Although researchers may "take great pains not to adopt the language of causality," they often fall "into this practice . . . when exploring 'neighborhood effects'" (Furstenberg et al 1999, 20). Researchers, policy makers, community activists, employers, and the public at large must resist the temptation of oversimplistic explanation. The following chapters aim to debunk the discursively produced—yet flawed—logic of neighborhood effects and examine the wider forces of identity formation and cultural exclusion.

In Search of Causality

The neighborhood effects literature has demonstrated beyond reasonable doubt is that a statistical correlation between neighborhood context and individual labor market situation exists. However, it is a difficult task to empirically show that representations of neighborhoods are a constitutive force on the segmentation of inner-city workers. To link statistical correlations with processes of identity formation and cultural exclusion requires a shift in epistemological as well as methodological frameworks of analysis. In terms of epistemology (or how we construct valid knowledge), we should not assume that our own perceptions of norms and social order apply universally across spatial and cultural contexts (Foucault 1970), a fallacy of which much of the neighborhood effects research is guilty. The difficult task is to recognize how forces of spatial and cultural representation and ideological differences between communities include and exclude human beings. To pursue this task we need to resort to appropriate methodologies (or research techniques). "Research on inclusion/exclusion . . . may require an appreciation of other worldviews, which is most likely to come from ethnography and participant observation" (Sibley 1998, 120).

The appreciation of other worldviews requires that we address normative issues in the search for an explanation for the labor market situation of inner-city youths. In particular, spatial variations between value systems, social protocols, and cultural coding schemes must be recognized. For example, the perception of acceptable and pathological behavior, the contents of labor market identity, and external and internal representations differ between neighborhoods. Without addressing these issues, it would be difficult to understand the situation of young people in the labor market. Quantitative neighborhood effects models that summarize social processes across multiple and diverse neighborhoods are inappropriate methods to answer questions of spatial and cultural particularity. These models rely on the assumption that certain universal principles of social behavior underlie all neighborhood-based social systems. However, social life is culturally and spatially fragmented, and it is difficult, if not impossible, to abstract social processes to such a degree that they can be summarized under the umbrella of a universal theory and still be meaningful to understand the concrete circumstances of individual neighborhoods. Thus, our understanding of neighborhood-based processes must be empirically grounded in the cultural context of the particular neighborhood.

Despite these limitations, quantitative models are not without value. They can measure, for example, the general correlation between the socioeconomic status of neighborhoods and the labor market situation of young people. Although this correlation says nothing about causality, it does generate new questions about the processes that create this relationship and thereby set the stage for further analysis. Quantitative models that examine a given city can also identify individual neighborhoods in which local socioeconomic circumstances are associated with high (or low) levels of labor market marginality among young people. But the quantitative modeling of neighborhood effects reaches its limit with the realization that cultural practices and ideological structures are spatially and culturally contingent.

To examine causal processes that operate in individual neighborhoods, an ethnographic method must be used. Ethnographic studies of particular neighborhoods are sensitive to the fundamental spatial and cultural contingency of social life on the supply side of the labor market, and they aim to understand the intrinsic logic of local social processes. Of course, this method has its own limitations, namely that it does not offer insights that are necessarily representative of other neighborhoods but that may be particular to the neighborhood under examination. The differences between generalizability and particularity, and between description and revealing causality, constitute a wide epistemological and methodological gap between quantitative and ethnographic research approaches (Sayer 1984).

In the following chapters I attempt to bridge this gap between the two research traditions. Chapter 2 presents a statistical analysis of census data that quantifies the relationship between the labor market prospects of young people and neighborhood characteristics in San Antonio. Chapters 3, 4, and 5 explain some of these relationships by engaging in an ethnographic study of two distinct neighborhoods. The differences between the quantitative and ethnographic approaches are fundamental. The quantitative analysis relies on the language of mathematics and follows a rigid positivist method. The availability of information contained in the census sets the parameters for this study. The ethnographic study, on the other hand, is exploratory in nature and required a great deal of flexibility and the continuous modification of research questions, methods, and theoretical frameworks throughout the research process. Needless to say, it is impossible to completely eliminate the gap between the two approaches. However, some important bridges could be constructed that enable a cultural interpretation of the observed phenomenon of labor segmentation and neighborhood effects.

Notes

1. For reviews of spatial mismatch, see Bauder (2000), Holzer (1991), Ihlanfeldt and Sjoquist (1998), Kain (1992), Preston and McLafferty (1999).

2. Although Williams relates the bonding process associated with structures of feeling to trade unionism and class solidarity and to national and regional scales, he acknowledges in a 1984 interview with geographer Philip Cooke that "there are other bonding mechanisms in reality which are beyond national consciousness and class consciousness" (Williams 1989, 241). I suggest that place-based bonding processes create shared work identities on the neighborhood scale.

3. The literature addressing the spatial relationships between social practices, discourse, representation, and cultural meaning has exploded in recent years. For example, Agnew (1987), Lefebvre (1991), Sibley (1995), Smith (1989), and Soja (1996) all comment on these issues.

4. Critiques of environmental determinism have convincingly argued that physical infrastructure and design enhancements do not necessarily improve individual outcome (Bohl 2000). For an illustrative counterpoint, Vergara (1995) has documented the physical devastation of the "New American Ghetto."

5. For an overview of the taxonomizing effect of common-sense thinking and academic theorizing, see Foucault (1970).

6. Some neighborhood effects research addresses behavior such as child abuse and violent crime (e.g., Coulton et al 1999; Simcha-Fagan and Schwarz 1986). My critique, however, focuses on the manner in which the idea of neighborhood effects is applied to far more ambiguous behaviors such as teenage and

out-of-wedlock pregnancy, dropping out of school, welfare dependency, and labor market performance.

7. Thus, statistical neighborhood effect models measure the degree to which neighborhoods facilitate or constrain assimilation to dominant culture. They do not measure the movement of people on an absolute scale of normal and pathological behavior.

Chapter 2

Mapping the Labor Market

The Specter of the Alamo

The tourist industry celebrates San Antonio as a distinctly "Mexican" city. Vendors in El Mercado sell oversize sombreros, and loudspeakers blast the voice of Tejano singer Selena into the Paseo del Rio (or Riverwalk) where restaurants serve hot tacos and cold cervezas. Ironically, the center of this tourist mecca is the Alamo, Texas's "shrine" to Anglo-American bravery and martyrdom, and symbol of the defeat of Mexicans. The distortion of historical events at the Alamo and the manipulation of Anglo and Mexican identities associated with the "Shrine of Texas" symbolizes the extension of the battle of the Alamo into the everyday life of contemporary San Antonio (de Oliver 1996, 2001).

Originally, the Alamo was a Spanish mission called San Antonio de Valero. Under the custody of the Daughters of the Republic of Texas, a patriotic women's group that bought the property in 1905, the Spanish appearance of the site was modified to better illuminate the courageous sacrifices of Travis, Bowie, Bonham, and Crockett, the Anglo-American heroes of the 1836 battle. The numerous modifications included the restyling of the mission's façade in a fashion that does not reflect Spanish Baroque of the original design, but "urban Dutch architecture" (de Oliver 1996, 5). The courtyard of the Alamo is an entirely new construction with radial pathways converging at the Alamo chapel, following the design principles used by eighteenth-century European architects to construct the extravagant residences of absolutist monarchs. Inside the chapel a gallery glorifying the Anglo heroes of the Alamo displaced the original Franciscan appearance of the chapel (de Oliver 1996). The subordination of Latino identity in the presentation of the Alamo site reflects the subservient social and economic situation of Latinos in San Antonio today.

With a population of roughly 1.2 million,[1] San Antonio is the third largest city in Texas. Approximately half of this population identified itself as "Hispanic" in the 1990 census.[2] Unlike in most other regions of the United States, Hispanics have been in San Antonio for centuries, and the size of their population is largely the result of domestic in-migration and natural population increase rather than international migration. The deeply rooted Latino presence in San Antonio is distinguished by strong extended family ties, the common use of Spanish within the family, and the affirmation of the Catholic faith. Jankowski (1986) suggests that compared to cities like Los Angeles and Albuquerque, Latino youths in San Antonio are less assimilated to Anglo-American values and more committed to traditional lifestyles and norm.

Ever since the late nineteenth century, when Mexicans assumed subordinate labor positions in an agricultural economy, many Mexican-Americans have remained marginalized in the economy (Montejano 1987). Although Hispanics are represented in all segments of the current labor market, the majority of them have benefited little from the city's recent growth in prosperity and continue to occupy low socioeconomic status categories. According to the 1990 Census, 26.2 percent of San Antonio's Hispanic families lived in poverty compared to 8 percent of non-Hispanic families (U.S. Bureau of the Census 1997). The level of welfare benefits in San Antonio have been among the lowest in the nation, and many single mothers have suffered extreme material hardship (Edin and Lein 1997). Unemployment rates among Hispanics were 7 percent in 1990 but only 3.9 for non-Hispanics. Even among those workers who have jobs, many do not earn enough to rise above the poverty level. Per capita income averaged only $11,827 in metropolitan San Antonio, compared to $12,904 statewide and $14,420 nationally. Per capita income among San Antonio's Hispanics was only $7,309 versus $16,246 for non-Hispanics. According to local authorities, adult underemployment, as measured by earnings, is largely responsible for the high poverty rates (Abramson and Fix 1993; Capps 1996; Johnson et al 1983). Wage levels in San Antonio are so low that many young people and potential entry-level workers drop out of the formal workforce altogether or do not even bother looking for work in the first place (Konstam 1996a).

While wages and welfare benefits are low, San Antonio's economy has expanded rapidly since 1970. In October 1996, San Antonio's unemployment rate dropped to 3.7 percent, prompting the *San Antonio Ex-*

press News to report "S.A. sees lowest jobless rate ever" on its front page (Konstam 1996a). Although the high-tech and biotechnology industries, as well as finance, insurance, and real estate grew steadily during this time period, the majority of new jobs were in the service, retail, and tourism sectors. In the 1980s, for example, approximately 80 percent of newly created jobs were in the service and retail industry (Abramson and Fix 1993). In 1990, 36.7 percent of the metropolitan area's employment was in the service industry, and trade industries captured another 23.6 percent of the employed. Jobs in these industries are characterized by low wages, seasonal employment fluctuations, limited benefits, and part-time employment. Historically, a strong manufacturing sector offering higher-wage, unionized employment has been absent in San Antonio. Aesthetically offensive smoke-stack industries, with their expensive unionized labor force, were rejected in favor of the expansion of the local tourist industry (de Oliver 2001). In 1950, for instance, less than 12 percent of San Antonio's workforce was in manufacturing (Schwab 1992, 214). This figure was still below 15 percent in 1970, 1980, and 1990 (Cardenas et al 1993; U.S. Bureau of the Census 1997). Army and Air Force-related employment has also been in retreat as local military installations downsized during the 1990s or closed altogether.

During the period of economic expansion in the service and retail sectors, job growth nevertheless remained below population growth. This situation contributed to an overall increase in competition between workers for the available jobs and pushed down wages, particularly in the low-skill and entry-level occupations (Abramson and Fix 1993). High levels of part-time and temporary employment are also notable characteristics in these sectors. City politics during this period supported business investments that were not necessarily in the interest of the local Latino workforce. Support for large-scale developments, including Sea World (a marine animal park), Fiesta Texas (a theme park), or the Alamodome (an enclosed stadium), promoted tourism and generated largely low-skill and seasonal employment (Rosales 1999).

In addition to these structural disadvantages in the local economy, the Hispanics of San Antonio also confront ethnic discrimination. Cardenas et al (1993) observe:

> The high poverty rate is often attributed to the poorly diversified economic base, weak manufacturing sector, and relatively low wage scale. In addition, San Antonio has a long history of ethnically exclusionary labor and development policies, and differential treatment of the Mexican-American population. (160-161)

The origin of Latino labor market exclusion traces back to the era when Anglo-Americans battled Mexican forces at the Alamo. Since the independence of the state of Texas, when Mexicans were seen as cheap farm workers, external representations have continued to depict Latinos as inferior labor, channeling them into secondary occupations (Montejano 1987). Throughout the twentieth century, San Antonio has maintained a "caste-like" social order, which was enforced by economic, political, and legal structures that disenfranchised Latino residents (Jankowski 1986).

Social inequality and representations of Latinos as second-class citizens have not only produced a culturally segmented labor market, but they have also enforced residential segregation between Anglo, Latino, and socioeconomic status groups. In terms of ethnicity, Hispanics are over-represented in neighborhoods on the West Side and the South Side of San Antonio, while Anglo-whites are more likely to live in the northern parts of the city. A smaller African-American community clusters in distinct residential pockets surrounding the downtown core. An elaborate system of freeways, railways, and drainage ditches, and even the security perimeter of a jail, seal off minority neighborhoods from view and interaction with the tourist crowds that frequent the downtown core (de Oliver 2001).

In terms of socioeconomic status, San Antonio is one of the most segregated cities in the United States.[3] Since 1945, the city of San Antonio has annexed over 260 square miles of its surrounding territory. Rapid middle-class expansion occurs mainly on the northern fringe, while lower status groups concentrate in the urban core, most notably on the West Side and the South Side (Partnership for Hope 1993; Witte 1993). Henry Cisneros (1995, 1-2), former secretary of the federal Department of Housing and Urban Development (HUD) and also former mayor of San Antonio, describes the changes that occurred in the San Antonio neighborhood in which he grew up:

> Built in the 1920's, many of Prospect Hill's original residents were railroad workers of German descent. By the time I lived there though, Prospect Hill had become almost entirely Hispanic. Nearly every man on our block worked as an aircraft mechanic at nearby Kelly Air Force Base. . . . My parents still live in the same house in Prospect Hill, but the neighborhood has changed. Few young people live there; they're out in newer subdivisions that are closer to their jobs. . . . The neighborhood's average income is dropping, and many once-familiar stores closed their doors years ago.

There is little doubt that these neighborhood-based circumstances shape the labor market prospects of young people.

The battle of cultural domination and subordination continues in San Antonio. However, it is no longer fought with saber and guns on the grounds of the Alamo. Rather, it has been relocated to more subtle arenas, such as the labor market, in which neighborhood representation and exclusion are powerful weapons. The analysis below sets the stage for a detailed ethnographic study of neighborhood processes by examining the quantitative patterns of labor market marginalization among young people across San Antonio's neighborhoods.

Marginalization of Youths

Many young people who enter the labor market find only low-paying and often part-time or temporary employment. But not all young workers in these positions are marginalized; some, in fact, advance quickly into more favorable employment. Inequalities that result in greater labor market prospects for youths in some neighborhoods than in other neighborhoods are of particular interest in explaining the spatial and ethnic segmentation of labor.

Many indicators of labor market prospects for young people are measured in the U.S. Census. Instead of examining individual variables of employment characteristics, poverty, education, and school enrollment, I develop a comprehensive measure of labor market prospects that encompasses multiple variables and that reflects circumstances among young people who share a neighborhood. Fortunately, census data are available by census tract, which are geographical units designed to represent neighborhoods. The area under investigation covers the 226 census tracts of San Antonio and its suburbs.

A statistical data reduction procedure simplified the information of multiple census variables that represent individual aspects of labor market marginalization among eighteen- to twenty-four-year-old individuals (see Appendix: Principal Component Analysis for a detailed description of this analysis). This procedure derived a composite indicator of neighborhoods in which young people have low labor market prospects. I labeled this indicator "Marginalized Youths" because it reflects several characteristics that signify marginality: census tracts are differentiated according to the proportions of youths without high school diplomas, who live in poverty, and who are neither enrolled in school nor employed. In particular, the combination of not being in school and not being employed characterizes marginality. In the labor market, opportunities tend to be given to workers with educational credentials and work experience. Lacking both education and experience severely limits labor market prospects.

The statistical data reduction procedure separated neighborhoods labeled "Marginalized Youths" from neighborhoods in which young people tend to be either enrolled in school or employed. Apparently, neighborhoods in which young people are more likely to attend school or go to work are not associated with the characteristics of marginalized neighborhoods, such as high poverty rates among young people. This finding challenges the idea that the main problem associated with economic marginality among young people in San Antonio is low wages rather than nonemployment. Neighborhoods in which employed young people tend to live are not the same neighborhoods in which young members of poor families live. While low-wage employment may drive adult workers and their families into poverty (Abramson and Fix 1993; Capps 1995), young people who work tend to live in neighborhoods that are better off. This finding illustrates the different labor market situations between young people and adults.

In North American cities the characteristics of workers usually follow distinct residential patterns: employment figures, skill levels, and educational attainments tend to be lower in inner-city neighborhoods and higher in suburban areas and on the urban fringe. In San Antonio, one would expect similar city-suburb differences regarding youth labor market marginalization. To investigate spatial patterns, I projected the values of the indicator "Marginalized Youth" onto a census tract map of San Antonio (figure 2.1). The central business district (CBD) defines the center of the city, and Loop 410 can be used as a rough categorization scheme that separates the inner city from the outlying neighborhoods and the suburbs.[4] Neighborhoods in which young people are more likely to be marginalized are mainly in the inner city. On the West Side and South Side of San Antonio, the pattern of marginalized neighborhoods extends even beyond Loop 410. Only military bases (i.e., Kelly, Lackland, and Brooks Air Force Bases as well as Fort Sam Houston) disrupt this pattern. Low levels of youth marginalization are apparent on the North Side and particularly beyond Loop 410. Eleven of the 226 census tracts suffered from extreme youth marginalization (shaded darkest). These tracts are clustered in proximity to the CBD, especially on the West Side and the East Side. They are home to 4,756 predominantly Hispanic youths and young adults between the ages of 18 and 24. Among these young people almost two-thirds have not completed grade twelve, over half live in families with incomes below poverty status, and 46 percent are neither employed nor enrolled in school. This pattern confirms expectations that marginality concentrates in the inner city and in isolated, poverty-stricken areas.

Figure 2.1: Marginalized Youth in San Antonio

The neighborhood effects literature suggests that labor market prospects of young people are closely correlated with the demographic characteristics of the neighborhood. While I do not endorse the conventional explanation of the working of neighborhood effects—namely, that culturally pathological peer groups and adult neighbors cause individual failure—the statistical relationship between neighborhood context and youth labor market prospects is undeniable. The next analysis therefore seeks to measure and define this relationship. According to the neighborhood effects literature, three sets of neighborhood characteristics are associated with youth labor market outcomes: first, social behaviors, such as unwed parenthood and teenage pregnancy rates; second, the employment characteristics of local adults, such as mean income, occupational segmentation, and the occurrence of part-time and temporary employment; and third, ethnic and place-of-birth characteristics of neighborhood populations. To assess the relative importance of these three sets of characteristics, I use another statistical procedure that enables me to establish the statistical relationship between the indicator "Marginalized Youths" and a variety of variables that measure behavioral, employment, and ethnic characteristics of neighborhoods (see appendix: Regression Analysis for a detailed description of this analysis).

This analysis confirms that youth marginalization correlates with neighborhood-based conditions. First, social behavior on the neighborhood level is strongly correlated with youth marginalization. Social behavior was measured as the proportion of births by single mothers relative to births by married mothers, and as the proportion of a neighborhood's women who were never married. Although these measures represent only a small aspect of social behavior, the literature on neighborhood effects, drawing on culture-of-poverty and underclass ideas (Lewis 1965; Moynihan 1965; Wilson 1987), suggests that these particular behaviors play a key role in the procreation of neighborhood pathology. I am skeptical of this explanation, however, because it is by no means clear *how* the social behavior that is prevalent in a neighborhood causally relates to youth marginalization. An alternative explanation, for example, could be that labor market opportunities, educational resources, and support services are removed from neighborhoods in which perspectives of family life differ from "mainstream" convention and in which out-of-wedlock childbirth or teenage pregnancy are more likely than elsewhere. Teenage and unwed mothers may be disadvantaged in school and the labor market because they lack access to day care, not because they do not appreciate education or provide inferior labor. Although the statistical correlation between behavioral characteristics and marginalization exists, processes of causality remain ambiguous.

Second, I analyze the statistical relationship between employment characteristics of a neighborhood's adult population and youth marginalization. The measures that I use to represent a neighborhood's employment characteristics include mean household income, the proportion of temporary and part-time employment among workers, the relative distribution of the workforce across labor market segments, and non-employment rates. Mean household income has no significant effect on youth marginalization, and, surprisingly, adults working part-time are associated with improved labor market prospects for local youths.[5] The other measures, however, display an expected relationship to youth marginalization. The neighborhood rate of temporary employment has the largest effect of any of the measures. An only modest increase in temporary employment among a tract's adult workers corresponds to large declines in labor market prospects among young people. In addition, the proportion of adult workers in secondary occupations, such as private household workers, machine operators, helpers, and manual laborers, also correlates with higher levels of youth marginalization. Finally, if a neighborhood has a high nonemployment rate among adults, chances are that the indicator of youth marginalization is also high. These correlations can be interpreted in different ways. There could be a direct causal link between adult characteristics and youth marginality, as implied by conventional neighborhood effects research. Alternatively, both adult employment characteristics and youth labor market marginalization could result from underlying structural circumstances that disadvantage both adults and young people in the labor market, but that are not captured in the statistical data collected by the census. Neighborhood representations and processes of identity formation could be such underlying circumstances that are not available in statistical data sets.

Third, I examine the ethnic characteristics of neighborhoods for their correlation with youth marginalization. This relationship is not as strong as I expected. The proportion of foreign-born persons is relatively unimportant with regard to the labor market prospects of young people. Neighborhoods with high Hispanic representation experience somewhat increased levels of youth marginalization, but compared to other neighborhood characteristics, this correlation is weak. The data, however, may understate the influence of ethnic discrimination. The U.S. Census assesses ethnicity on the basis of origin and not on the degree of cultural assimilation. More acculturated Hispanics may reside in areas of lower levels of youth marginalization and experience less cultural discrimination. Less acculturated Hispanics, on the other hand, may cluster in tracts that are highly stigmatized and where young people suffer from this neighborhood stigma. Processes of cultural differentiation, rather than ethnicity, may be grounds for discrimination in the labor market and the educational system. Consequently, the proportion of Hispanics, as measured in the census, may not accurately capture cultural dissimilarity.

The main conclusion of this analysis is that local context is an important element in shaping the prospects of young people in the labor market. While a strong relationship exists between the marginalization of young people and neighborhood characteristics, I am highly critical of the explanation of causality advanced by the neighborhood effects literature. Statistical evidence requires a critical interpretation to establish causality between neighborhood circumstances and the labor market situation of youths. I argue that the influence of neighborhood conditions on youth labor market prospects is more complex than the statistical variables available in the census express. In particular, behavioral and labor market attributes are not only linked to local economic structures and the quality of schools but also to cultural representations. A rigorous qualitative analysis will focus on two neighborhoods to shed light on how place-based circumstances can influence the labor market prospects of young people.

Two Neighborhoods: Lanier Area and Palm Heights

While the above analysis reduced complex relationships of the labor market to a few quantifiable variables, it did reveal important common properties, correlations, and spatial patterns. Qualitative analysis, on the other hand, is more suitable for uncovering particular processes that produce the correlations and patterns revealed above. Since the quantitative relationships were framed by the neighborhood as the fundamental spatial unit, the qualitative analysis will employ the same spatial unit to unearth causal linkages. I use the results from the statistical analysis to select two case-study neighborhoods for an in-depth investigation: the Lanier area and Palm Heights (see appendix: Case Study Area Selection). Both areas are located in the inner city and on the Southwest side of San Antonio in relative proximity to the central business district (figure 2.2).

The census delineates neighborhoods through exact census tract boundaries. The statistical nature of the census demands that there is no ambiguity where one spatial unit ends and another one begins. For the purpose of an ethnographic study, however, it would be unwise to rigidly define neighborhood boundaries, because different groups and individuals have different perceptions of the spatial extent of the neighborhoods. My definitions of the two neighborhoods are therefore mere approximations of a commonly shared perspective among residents. Most residents, for instance, agree that their neighborhoods are organized around central institutions such as schools and community centers. I have used the most important of such institutions to give the two neighborhoods their names: Sidney Lanier High School and the Palm Heights Community Center.[6]

Figure 2.2: Lanier Area and Palm Heights

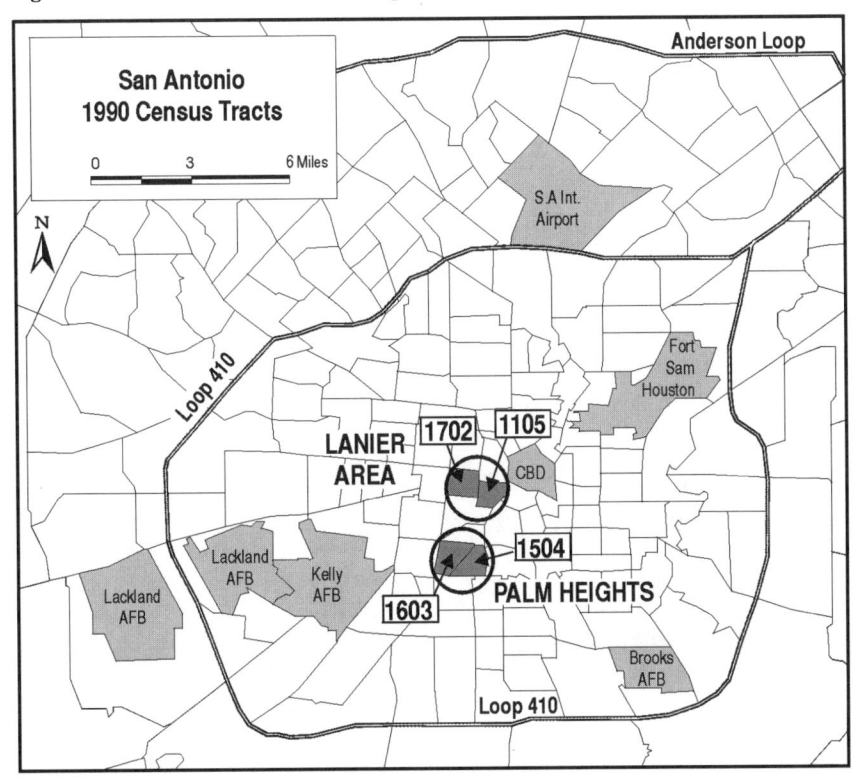

Despite the flexible interpretation of neighborhood boundaries, both the Lanier area and Palm Heights roughly coincide with two tracts (see figure 2.2). These tracts can be used to provide basic information about the demographic makeup of the two neighborhoods (see table 2.1). On the one hand, the two neighborhoods share several characteristics. Hispanics constituted between 94 and 98 percent of the population in all four tracts in 1990. The share of foreign-born population, however, was low, indicating that most Latino families in these two areas are not immigrant families. In both areas relatively few workers had professional and managerial jobs in the independent primary segment of the labor market (see appendix, table A.4, for the classification scheme).

Table 2.1: Lanier Area and Palm Heights Characteristics, 1990

	Lanier Census Tract		Palm Heights Census Tract	
	1105[2]	1702	1504	1602
Demographics, 16 and Older (%)				
Hispanic	97.4	97.2	95.6	94.0
Foreign born	14.2	19.8	13.2	11.3
Households with Non-English Speaker(s)	42.8	30.2	13.2	14.3
Births by Mothers under 17[1]	5.4	8.2	6.7	1.9
Births by Single Mothers[1]	52.3	30.6	22.5	15.4
Poverty Level	61.6	37.0	24.2	18.2
Household Income ($)	3,635	8,024	15,316	15,837
Persons not working	67.8	53.5	42.3	41.4
Workers in Independent Primary Occupations	2.9	13.2	13.7	16.2
Workers in Subordinate Primary Occupations	32.4	47.0	54.7	50.7
Workers in Secondary Occupations	64.7	39.8	31.7	33.1
Persons with Part-Time Jobs	8.2	8.1	11.4	10.2
Persons with Temporary Jobs	12.0	11.4	14.0	12.3
Youth Careers, 18 to 24 (%)				
At Work	31.5	50.0	70.6	50.9
Enrolled in School	8.0	26.8	34.9	45.9
Neither Enrolled nor at Work	62.0	37.3	18.3	25.0
Not Completed Grade 12	72.1	61.6	28.3	51.8
Total Number of Youths	387	662	438	599

Sources: U.S. Bureau of the Census (1991); San Antonio Metropolitan Health District (1994)
[1] Data for 1993
[2] Contains the Alazan-Apache Housing Courts

Despite these commonalities, there are important differences. In the Lanier area, more households had at least one non-English-speaking member, births by single mothers was more frequent, poverty levels were higher, and household incomes were lower than in Palm Heights. In addition, a larger share of Lanier area workers were employed in secondary occupations, while most Palm Heights workers had "subordinate primary" occupations such as sales, administrative support, or transportation-related occupations. Most significantly, however, the Lanier area contains the Alazan-Apache Housing Courts, which is San Antonio's largest public housing project. In 1996, it had 1,015 housing units for low-income families. Mean household income in Tract 1105—the site of the housing project—was only $3,636 in 1990, compared to $33,648 countywide. The poverty rate in this tract was 61.6 percent, compared to 26.2 percent in Bexar County. Sixty-eight percent of this tract's population was not working.[7]

In addition to these differences, the two neighborhoods varied with respect to youth labor market prospects. Compared to Palm Heights, few young people in the Lanier area had jobs or were enrolled in school. More young people in the Lanier area did not complete grade twelve and, most importantly, more were neither enrolled in school nor working.[8] All together, 1,049 youths lived in the Lanier area and 1,037 lived in Palm Heights (see table 2.1). The mapping of the indicator Marginalized Youths (see figure 2.1) illustrates the different situations of the two neighborhoods. The Lanier area is in the center of several census tracts on the West Side in which levels of marginalization are extremely high. The two census tracts that cover Palm Heights, on the other hand, stand out as having low levels of marginalization compared to the surrounding tracts.

Table 2.2: Employment by Zip Code, 1990

Area	Firms	Employees
Lanier Area (78207)		
Retail	206	2,403
Service	189	1,467
Manufacturing	49	n.a.
Palm Heights (78225)		
Retail	50	529
Service	30	171
Manufacturing	11	n.a.
Central Business District (78205)		
Retail	303	4,831
Service	652	10,558
Manufacturing	16	n.a.

Source: U.S. Bureau of the Census (1992)

The location of the two neighborhoods relative to job opportunities is unlikely to affect employment differences of youths between the two areas. The zip code containing the Lanier area offers more jobs in the retail and service industry than the zip code containing Palm Heights (see table 2.2). In addition, the Lanier area zip code has more manufacturing firms that Palm Height's zip code. Furthermore, the Lanier area is located closer, and presumably has easier access, to a large pool of entry-level service and retail jobs in San Antonio's central business district. The presence of large numbers of employment opportunities dismantles the argument that spatial mismatch could disadvantage young people in the Lanier area compared to young people in Palm Heights. Other forces must be at work to produce the high level of observed youth marginalization.

Between September and December 1996, I conducted interviews with twenty-one young persons between the ages of sixteen and twenty-five who lived in the Lanier area and with eight youths in the same age group who lived in Palm Heights (see appendix: Method of Qualitative Analysis). A second set of interviews included representatives of six community organizations in the Lanier area, three organizations in Palm Heights, and eight organizations that serviced both areas. I collected additional information through numerous informal conversations, volunteering in a youth group, and participating in community events. During the months of conducting fieldwork, I lived in an apartment complex on the perimeter of the Lanier area, where I shopped, frequented restaurants, and spent much of my free time.

Notes

Portions of this chapter are based on a paper that I coauthored with Bob Sharpe.

1. The statistics on San Antonio refer to Bexar County, which contains the city of San Antonio and its suburbs. San Antonio proper had a population of 935,927 in 1990.

2. "Hispanic" is the category used in the U.S. Census, and I use this terminology when I refer to census data. Residents often used the terms "Latino" and "Latina" when they referred to themselves, and I use these labels to reflect internal identities. Other labels include "Chicano" and "Chicana," "Mexican," or "Mexican American." To be consistent I use the former two terms in the text, unless other labels are used in direct quotes.

3. Abramson et al (1995) used the isolation index to measure the extent to which the poor are exposed in their neighborhood only to members of their own group. With an isolation index of 30.0 percent, the poor in San Antonio have the sixth highest isolation index among the 100 largest U.S. cities. This compares to a national average isolation index of 21.0 percent. Further, this index changed negligibly between 1970 and 1990.

4. Although the municipal boundaries of the city of San Antonio stretch beyond Loop 410, I use this roadway as a convenient marker of city-suburb differentiation that reflects the perception of many residents.

5. The neighborhood effects literature provides little guidance for explaining this relationship. I can only speculate that part-time workers may have more time to supervise young people, or that they may be more likely to live in well-off

neighborhoods where households do not depend on dual full-time employment incomes.

6. Not all residents labeled their neighborhood "Lanier" and "Palm Heights." Other names that identify locations in the Lanier area, for example, are La Tripa, El Con, Ghost Town, or simply "the West Side." For the sake of consistency, however, I use the names Lanier and Palm Heights throughout the text.

7. Anderson (1999, 16-34) attempts to describe fundamental differences in behavior, rules, and codes between neighborhoods along Philadelphia's Germantown Avenue. The risk of such description is, of course, to create stereotypes of "authentic" neighborhood culture.

8. Countywide roughly 30 percent of youths 18-24 have not completed grade 12 and about 20 percent are neither enrolled in school nor employed.

Chapter 3

Representing Neighborhoods: Views from the Inside and the Outside

Youths and Identities

The geographical context of the neighborhood is more than a mere container that configures a set of independent labor market variables. Rather, neighborhoods frame important processes of identity formation and cultural representation. These processes in turn shape labor market expectations and career prospects of young people. I distinguish between two kinds of processes. First, "internal" processes of identity formation create a sense of belonging that relates the lived experiences of the everyday to the context of the neighborhood. Second, "external" representations enable people to differentiate between "us and them." This distinction is important because it identifies two distinct (yet related) processes that reflect different labor market identities.

Internal processes of identity formation are deeply entrenched in the cultural context of the neighborhood. For many of us, the sense of belonging is tied to local territory. Residents are emotionally connected to their neighborhood through personal networks; they are accustomed to the built environment, and they are familiar with local social practices and conventions. Youth groups often express their identities through territorial claims of city blocks, neighborhoods, or West and East Sides (Bourgois 1995; Ley 1974; Shuttles 1968). Sometimes, this sense of belonging is formalized through involvement with neighborhood-based organizations, local interest groups, and community events. These internal neighborhood identities can shape the work expectations of young people and influence career-related decision making.

External representations, on the other hand, express neighborhood identities that are imposed from the outside (Anderson 1991; Smith

1989). They tend to suggest cultural homogeneity and conceal ethnic, behavioral, and material differences within a neighborhood. Media reports, for example, have fed a popular discourse that represents minority neighborhoods in our inner cities as cultural wastelands in which civility has given way to social disorder and street violence (Wilson 2001). In a similar vein, these images of inner-city neighborhoods also evoke expectations of skill deficiencies, low education, poor work ethic, gendered work roles, and occupational incompetence among residents. If these stereotypes are shared among employers, educators, and administrators of community-based organizations, then they can have severe and long-lasting impacts on the employment prospects of young residents.

In comparison to adults, young people are probably more affected by the cultural context of their neighborhood because they are relatively confined to the local environment. In the Lanier area, spatial confinement is particularly acute. Although a stream of images reaches Lanier area youths through national and international media channels (TV, movies, music, magazines), personal interaction takes place on the local scale. If interaction with nonlocals occurs, it is usually in formal settings, such as in school, and involves authoritative figures such as teachers or police officers with whom few youths immediately identify. Few youths travel outside their neighborhood in their daily paths of movement.

For some young women, time limitations can be severe in light of simultaneous household, educational, work, and parenting responsibilities (Kwan 1999). A teenage mother, for example, quit her job six miles away from her home because the daily commute was too long: "I was pregnant and had to take the bus. I had to get up at 5:00 in the morning take three buses to get to work, take three buses to get back home." She therefore switched to a job within walking distance. Only one of the Lanier area interviewees owns an automobile. Most other youths depend on rides from family or friends to visit places beyond the boundaries of their neighborhood.

Fear of harassment is an additional barrier that prevents many youths from crossing neighborhood boundaries (Bourgois 1995; Davis 1990; Valentine 1992). A seventeen-year-old mother and Lanier area resident would not walk beyond Commerce and Trinity Streets on the perimeter of the neighborhood because of fear of sexual harassment and violence. Since she does not have a car, she stays within a few-block radius of her house or asks for a ride from family members or friends. Many young men reported being surveilled and confronted by police when they hang out in downtown tourist and shopping areas. In addition, most youths are simply unfamiliar with the cultural contexts of other areas. A seventeen-year-old student said that she does not visit the North Side of San Antonio, "cause I don't know this area."

With relatively little exposure to places outside the Lanier area, young people frame their identities primarily in reference to the neighborhood context which is immediately accessible to them. A Lanier area youth counselor suggests that the activity spaces create locally particular worldviews. He states: "They [youths] think the Lanier area is their world. To many kids, the reality is the neighborhood. They know the North Side, but the North Side is not their reality." The immediate neighborhood context provides the spatial reference for the formation of work identities.

Lanier Area Representations

The residents of the Lanier area see themselves as diligent and industrious workers. They also embrace distinct ethnicity- and gender-specific ideas of work. Many men, for instance, aspire to manual labor jobs, mechanical occupations, and construction work, while women often assume roles in personal services and low-paying manufacturing occupations. These labor market expectations no doubt influence Lanier area youths in their career decisions. One needs to be cautious, however, not to overgeneralize the effect of these local labor market identities on young people. Individuals still make their own career choices and develop particular work expectations.

Of the twenty-one young interviewees, four had full-time and six had part-time jobs at the time of the interview. Another seven were looking for work. Only three interviewees had never worked before. Six of the twelve women who were interviewed were employed. They included a babysitter, a landscaper, a City-Year volunteer, an amusement park worker, and two secretaries. Four of the nine male interviewees had jobs, including a health worker, a busboy, a fast-food server, and a City-Year volunteer. Only one young man and two young women made above $7.00 per hour.[1] Most young interviewees had previous jobs as restaurant, hotel, and cafeteria workers; cleaners; retailers; babysitters; telemarketers; and temporary workers. Future work aspirations mentioned repeatedly by young women included nursing, health care occupations, and social work. Male respondents aspired to military service, restaurant, and construction work. Notably, many young men and women were satisfied with secondary work in the restaurant, hotel, and retail business.

The content of local work identity is brought to the forefront when young people challenge local expectations. For instance, the members of one particular youth group aspire to manual work in the construction and service industry, and they negate the value of education to achieve their goals. The group rejects other youths who embrace different work identities. A sixteen-year-old male high school student complains that his college aspirations conflict with the norms that prevail in this group: "A lot of my friends are dropouts. They call me a sellout because I'm moving on to better things."

Local work identities also reflect gender-particular roles. A twenty-one-year-old office assistant, for example, disagrees with the idea that, as a Latina woman, she should assume minimal responsibility at her job. She argues with her father about whether or not professional jobs are better suited for privileged white women:

> My dad kinda makes fun of me: "You think you're a white girl." "No dad." "Then don't act like one." You know what I mean? I told him: "You know dad, I found a 50,000 dollar mistake [in the bookkeeping at her work]." I thought he'd be proud of me. And he got after me and said: "Don't tell anyone cause it's gonna be your fault." And I was like: "Dad, calm down, I already told [my boss]." But my dad would jump on me [. . .] "you don't want to take those responsibilities."

The director of a job-training agency argues that rigid career expectations for Latina women prevent local women from penetrating male-dominated occupations. Instead, many Latina women remain locked in cleaning and caretaking occupations:

> San Antonio in the past seven, eight, nine years has tried about four, five times to start a program for women in non-traditional work. These are the ladies that work on the expressways, the highway patrol police — ten, twelve bucks an hour — construction workers, all this stuff. The program has not been successful. They were [implemented] by good, strong organizations, but [a] lot of women, with welding [jobs] and stuff like that: "Ah, I'm not gonna do that. It's not what you're supposed to do." And some of these women are second or third generation here in Texas. Here, they all go into traditional occupations: teaching, daycare.

Although gender and ethnic typecasting are not unique to the Lanier area, they are mediated through local contexts. The neighborhood is the mechanism through which work identities and labor market segmentation are reproduced. The social context of the Lanier area sets a general standard for career success and failure that differs from other neighborhoods in

San Antonio. A local community activist emphasizes that perceptions of what constitutes desirable career paths within the local community differ from other parts of San Antonio. He explains that among Lanier area residents, an occupation such as "flipping hamburgers" is often considered a successful career. On the North Side, however, flipping hamburgers is regarded a failure. In support of this observation, a young interviewee declared that "dishwasher" is his long-term career goal. Within the particular context of the Lanier area, this career goal meets expectations. In the eyes of the outsider, however, his aspiration is not a valid career goal.

Local idiosyncrasies are an important reason for neighborhoods to exert a distinct force on work identity and occupational aspirations. For instance, many Lanier area youths base their perceptions of desirable and attainable occupations on the industries and businesses that are visible within the neighborhood. Seven of the twelve young Lanier area women I interviewed aspired to health-care-related occupations, such as nursing, nursing assistant, or health-care worker. While these aspirations reflect a female caretaker role in general, they respond in particular to the presence of the Santa Rosa hospital complex and the University Health Center at the perimeter of the Lanier area, which makes health-care-related job opportunities seemingly plentiful and accessible. Most of the training Santa Rosa administers, however, is for patient-care assistant, an occupation that paid an hourly wage of only $6.50-$7.50 at the time I conducted this survey (Konstam 1996b). Among male youths, the presence of restaurants, repair shops, manufacturing firms, and nearby Kelly Air Force Base influence career aspirations.

From an outsider perspective, the Lanier area symbolizes the collective negative characteristics of its residents. External representations portray Lanier area residents as dirty, aggressive, and unreliable "Mexicans" (although most of these families have been in the United States for decades or generations). Representations of Lanier area youths evoke not only images of street violence and drug trafficking but also disrespect for authority, confrontational attitudes toward adults, and a general lack of work ethic. The Alazan-Apache Housing Courts—the ultimate symbol of labor market failure—magnifies the negative image of the neighborhood. These perceptions of the Lanier area represent residents as undesirable workers. Young men and women are especially burdened with these negative representations.

Even Latinos living in other neighborhoods dissociate themselves from the negative images of the Lanier area. When a local youth worker—a Latino himself—first moved to the Lanier area, his family criticized his intention because of the low status of the Lanier area. He recalls:

> When I came to San Antonio I lived with my sister for a while. I said "I'm moving to the West Side." And she said "To the West Side, why do you want to live on the West Side?" "I need to live in the community where I serve young people." And she couldn't understand, because in her mind, the West Side was a dirty place, those were Mexicans and "I don't want to be with those Mexicans. I'm a Mexican-American because I was born here."

A college placement specialist, who is also Latino, harbored similar stereotypes when he began working in the Lanier area. He remembers his expectation of neighborhood crime and violence:

> When I first came over here I thought: "Oh my god, do I really want this job." I come from the north-west side of town and I was all scared that my car is gonna be vandalized and broken into, stolen. I was thinking the worst. But once I was here for a while . . . it's not as bad as people say it is, you know. But that's what I was thinking.

External representations also include expectations of labor market performance. Both the young people and the institutional administrators I interviewed suggested that employers and educators routinely reject applicants from the Lanier area because of the perception that they are unreliable and incompetent. A training provider expresses his frustration with an employer who dismissed his clients after learning that they live in the Lanier area:

> When people say "West Side" they automatically think you're poor, you belong to gangs and you're on welfare. So the stigma is there. I have two students that graduated from here. We trained them up front, we sent them to the hospital to get on-the-job training. And the [employer] told me they don't like to get welfare [-to-work employees], because they know they're not dependable.

The entire neighborhood suffers from this negative label, even the residents who are not on welfare and who do not live in the public housing courts.

Many Lanier area youths have a "West Side" dialect that reveals their ethnic and residential origin. This dialect, in combination with other physical markers such as clothing and hairstyles, symbolizes a cultural capital that is ill-suited for many occupations. A youth employment counselor explains that employers often evaluate young job applicants based on their speech and appearance:

Counselor: A lot of them [youths] have what they call their West Side tone and, you know, the way they speak. And I think, if I was an employer, would I want to hire them?

H. B.: So they are labeled by the employers?

Counselor: Right. West Siders talk different and they dress different. And it's normal for them to dress like that here. So, yeah, they have difficulties finding jobs. Their aspirations may be high; eventually they come down.

Many teachers and educational administrators hold similar stereotypes of Lanier area youths. The executive director of a skill training institution suggests that negative perceptions of the Lanier area channel local youths into secondary occupations. She observes in particular reference to the stereotypes of the Lanier area: "One of the biggest negatives I see [in San Antonio's school system]: because a student comes from a poor area, maybe they ought to go vocational. And that's not true." The external stigma attached to the Lanier area restricts educational opportunities and subsequently narrows the range of career paths available to young residents.

Stereotypes can depress the self esteem of young people. An eighteen-year-old college student describes the negative consequences of ethnic stereotyping that caused many of her peers to drop out of high school. She explains: "If someone says to me 'Oh boy, you're not gonna make it because [. . .] you're Hispanic, how should we expect you to accomplish.' And they [her Latino peers] would go like: 'Oh you know, you're right' and so they give up." The stereotypes of the Lanier area affect residents in a similar manner. A young interviewee believes that once "you live in the barrio, you're trapped" because neighborhood stigma denies upward mobility to its residents. A youth group coordinator confirms that the cultural representation of the Lanier area depresses aspirations among young residents to pursue careers in upper labor market segments:

[Lanier area youth] feel that they are inferior. I mean just be blunt. They feel that they are inferior . . . that the people that are in the northern part [of San Antonio] are more likely to have a better job. A job that won't be manual where they be working and using their muscles and working in the sun—where[as] they themselves would have to be clerks or cleaners.

Thus, Lanier area youths internalize external representations, turning stereotypes into self-fulfilling prophecies.

While internal and external representations of culture and neighborhood are conceptually distinct processes, they do construct similar work identities for young people, endorsing occupations with low wages, few benefits, and little job stability. However, the problem is that conceptions of work and marginality are spatially and culturally contingent. Internal and external perspectives reflect different standards of what constitutes acceptable work and labor market success. Internal representations position the career aspirations of Lanier area youths within the range of acceptability, while external representations situate them on the margin of the labor market. The work identities of Lanier area youths are not naturally marginal: dishwasher or nursing assistant are not necessarily "bad" careers. Only in reference to better-paid careers such as restaurant manager or surgeon do these careers become secondary. The idea that Lanier area residents are marginal outcasts reflects an outsider perspective.

My interview with the executive director of a local youth center exemplifies the importance of a cultural reference system in which ideas of success and failure are rooted. While she recognizes that local youths are committed workers, she applies an outsider perspective that places these youths on the margin of the labor market:

> The kids that we have here, I can tell you, are very hard workers and I think that they would do well in the manual jobs because they are not afraid, you know, to get out there and work hard. And some of them will probably end up there because they already feel that they are so poor in terms of academics that they wouldn't be able to get a professional job. And you know, there are a few that are quick learners but it couldn't be anything that involves too much of their own thinking and decision making because they're so used to being told, you know, what to do, what to say and what to write. Very few can think on their own.

Lanier area youths are, of course, not intellectually inferior to young people who live elsewhere. Rather, they embrace different ideas of career success than the executive director. For the director, the few "quick learners" are young persons who have adopted ideas of work and success that are similar to her own. The others, who can't "think on their own," are those who continue to embrace identities that reflect perspectives which are prevalent in the Lanier area. The respondent's outsider position shapes her impression of Lanier area youths as marginal.

Palm Heights Representations

Similar to people in the Lanier area, the residents of Palm Heights see themselves as workers who do not shy away from hard labor. However, according to census statistics, they are slightly less likely to work in manual labor and personal service occupations at the bottom of the labor hierarchy than Lanier area residents. They are more likely to work in middle segment occupations in technical and administrative jobs.

Young people in Palm Heights aspire to professional employment. All respondents were still in school at the time of the interviews; one female respondent had a part-time job in a clothing store and one young man was a part-time computer technician. Three respondents were looking for part-time work, and only three interviewees had previous job experience. Future aspirations among all interviewees focused on college education and professional occupations, such as being an engineer, pediatrician, zoologist, and computer specialist. These professional work identities differ strongly from the secondary work identities of young people in the Lanier area.

Unlike in the Lanier area, young people in Palm Heights do not reproduce the work identities of their parents and neighbors. By the same token, the local community does not perceive the professional career aspirations of young people as a threat to its own identity. Youths who aspire to careers in the primary segment of the labor market do not confront resistance from peers, family members, or local adults. Rather, the interviews revealed that parents and the community generally offer support and encouragement to young people who pursue college educations and professional careers.

An important aspect of the career development of local youths is that the Palm Heights community represents itself as a neighborhood that is poor but not at the bottom of the hierarchy of San Antonio communities. The Palm Heights community constructs an internal identity in opposition to the West Side communities (including the Lanier area), which are perceived to be lower in social status. According to a community activist who grew up in the Lanier area, but who now lives in Palm Heights, the image of Palm Heights is that of a poor but "honest" neighborhood. In comparison, people in the Lanier area are not only poor, they are also "on welfare and government assistance." Thus Palm Heights residents distance themselves from the cultural dysfunction that they perceive to exist on the West Side.

The presence of several public housing projects on the West Side, including the Alazan-Apache Housing Courts in the Lanier area, aggravate the negative images of West Side neighborhoods, while the absence of public housing in Palm Heights enables local residents to perceive their own neighborhood as equipped with the cultural capital to join the middle-class communities of San Antonio. The strategy of upward mobility includes the acquisition of middle-class occupational identities. Young members of the community, in particular, are encouraged to pursue careers in professional occupations and the primary segment of the labor market. Flipping hamburgers is not seen as a career through which the goal of upward mobility can be achieved.

An additional aspect that influences the formation of work identities among young people in Palm Heights is that their activity spaces are less constrained than those of Lanier area youths. Young people in Palm Heights frequently attend events in other neighborhoods and maintain peer networks that cross neighborhood boundaries or leapfrog into other parts of San Antonio. For instance, Palm Heights youths play basketball not only at the Palm Heights Community Center but also at other community centers throughout San Antonio. Some young interviewees have friends who live on the North Side of San Antonio. The spatial integration of social networks and the wider activity spaces expose Palm Heights youths to cultural contexts beyond their own neighborhood. This exposure breaks the cultural isolation of youths and widens their perspectives of work and career.

External representations also differ from the Lanier area. Most importantly, the absence of a large public housing project immediately improves the image of the neighborhood to outsiders. Palm Heights is seen as a decent neighborhood where families earn their living rather than living off public transfer payments, like welfare families do. Growing up in these decent families—so goes the outside perception—young people learn the value of hard work early on. As diligent and reliable workers, they possess the work ethic necessary to succeed in the labor market. These external perceptions affect the career development of Palm Heights youths positively, as employers, educators, and the staff of community-based institutions draw on these images when they judge and interact with local youths. For instance, in contrast to Lanier area youths, the youths I interviewed in Palm Heights consistently reported that teachers encouraged them to achieve in the school system and to pursue their career aspirations. This encouragement is motivated by the belief that Palm Heights youths are not "dysfunctional" but have the potential to acquire middle-class status.

Although young people in Palm Heights are just as "Latino" as their Lanier area counterparts, they apparently do not suffer in the school system and the labor market from negative stereotyping to the same degree as Lanier area youths do. This difference in treatment suggests that ethnicity alone does not account for the external perception of young people. Rather, neighborhood of residence emerges as a defining element. Average household incomes are higher in Palm Heights, poverty levels are lower, and more workers in Palm Heights are employed in mid-level administrative, technical, and transportation related occupations. However, relative to the full spectrum of San Antonio neighborhoods, both the Lanier area and Palm Heights are of low socioeconomic status. This situation suggests that local particularities, especially the presence of a public housing project, shape outsiders' perceptions of the neighborhood, which in turn shapes the futures of the young people.

Conclusion

The evidence presented in Chapter 2 and in this chapter derives from different research traditions, and, accordingly, differs in both the content and the manner in which it is presented. In Chapter 2, I analyzed the quantifiable relationship between neighborhood context and the labor market prospects of young people on the wider scale of San Antonio. In this chapter, I examined qualitative issues of representation in two particular neighborhoods. Now I want to present examples of how the findings from the two chapters complement each other. There are several bridges that span across the gap between the quantitative and ethnographic research approaches.

For example, the statistical analysis revealed a correlation between the proportion of workers in secondary occupations and high levels of youth labor market marginalization. In light of work identities that are commonly shared by young people and adults, this correlation should come as no surprise. Internal work identities circulate within neighborhoods, and the production of these identities involves both adult residents and young people. The circularity of processes of identity formation explain the observed correlation between the labor market characteristics of adults and young people. Similarly, external representations of the Lanier area categorically label both young people and adults as inferior workers. A valid question therefore is whether the correlation between high rates of temporary employment among local adults and high levels of youth marginality in a neighborhood relates to external stereotypes and discrimination that affect young and older residents alike.

The weak statistical relationship between ethnicity and youth labor market prospects can be explained by the ambiguity of the category "Hispanic." This category treats all persons with origins in Spanish-speaking countries alike. It does not express complex cultural differences within the Latino population or between neighborhood-based communities. Cultural identity may instead be reflected by different perspectives of family and child rearing. These perspectives are captured by statistical variables, such as a neighborhood's proportion of single mothers and women who were never married, which do correlate with youth marginalization.

Besides illustrating the relationships uncovered in Chapter 2, this chapter also produced findings that relate to more general issues of work identity, labor segmentation, and neighborhood representations. The work identities of Lanier area youths parallel the aspirations of working-class lads in England described a quarter century ago by Willis (1977). This consistency across time and space suggests that the work identities of Lanier youths are not entirely unique, but follow a general process of identity formation that reproduces the social order of capitalist society through the cultural segmentation of labor. Representations of the neighborhood are a critical mechanism in the production of labor market identity. Internal perceptions of what constitutes success and failure in the labor market are produced and shared within local social contexts. External representations of youths' labor market identity are also place based. The perspective that neighborhood of residence is a constitutive element in the segmentation of labor extends contemporary labor market theory. This perspective stresses the importance of cultural representations of place.

The problem for Lanier area youths reaches deeper than simply "choosing" the wrong identity. There is no logical reason for Lanier area labor to be so grossly undervalued in the labor market. Newman (1999, 86) correctly observes that "We carry around in our heads a rough tally that tells us what kinds of jobs are worthy of respect and what kinds are to be disdained." What is problematic for Lanier area youths is that their tally differs from that used outside the Lanier area, including Palm Heights.

Segmentation theory has shown that pay scales, job security, and benefits are not pure market creations. Rather, labor reward structures are shaped by discursive practices that revolve around cultural struggles of domination and are often camouflaged as "economic" processes in neoliberal rhetoric (Gibson-Graham 1996; Sassen 1994). The labor market itself is a mechanism for social distinction. Yet, the common perception that the economy is a "natural" process, and that the invisible hand of the market compensates workers "appropriately" for their labor legitimizes and reinforces this mechanism.

From this perspective, it is not coincidental that the work identities of young people in the Lanier area correspond to occupations that are allocated to the lower end of the labor market spectrum. Washing dishes and flipping hamburgers are not "bad" jobs per se, but the skills and work tasks are undervalued in comparison to a restaurant manager. The residential segregation of people and the devaluation of their labor are part of a wider process of cultural exclusion that marginalizes young Lanier area residents. The external representation of Lanier area youths as inferior labor diminishes the status of Lanier area residents and simultaneously secures the supply of cheap labor to the economy. Place of residence is a defining factor in this project.

Practices of community-based institutions reveal how external and internal representations, perceptions of marginality, and work identity interact with each other to produce different labor market outcomes for young people in the Lanier area and Palm Heights. Chapter 4 focuses on these institutional practices.

Notes

1. For reference, in August 1996 the Clinton Administration raised the federal minimum wage from $4.25 to $5.15.

Chapter 4

"You're Good with Your Hands, Why Don't You Become an Auto Mechanic?"

Institutions and Young Careers

The neighborhood of residence is a constitutive force on the development of youths. However, we must be cautious not to endorse the logic of the neighborhood effects idea that blames inner-city communities for their own marginality. In this chapter I suggest that institutions participate in the ideological colonization of inner-city minority neighborhoods like the Lanier area. In some instances, institutional practices even depress youths' career expectations and discourage achievement in the educational system and the labor market. It is important to note that these effects are not the intention of the institutions which I interviewed. Rather, institutions operate within the popular discourse of the inner city and make normative choices that can inadvertently disadvantage some youths.

The internal and external representations of neighborhoods are intimately related to institutional practices and the treatment of local youths. Community-based institutions do not deliver services in an objective and value-free manner. Rather, they are guided by ideological presumptions of what constitutes labor market achievement. These presumptions define the "proper" sequence of life priorities and offer ideology-laden explanations for success and failure. Institutional prescriptions for assisting young clients in building successful careers follow the principles laid out by the particular worldviews of administrators and the image these administrators have of the neighborhoods they serve. Few administrators accept the self-defined career identities that youths carve out for themselves. Some administrators understand that internal and external representations confuse youths' self-identities, and they attempt to mediate between the two positions. Many administrators, however, embrace external perspectives of cultural norms, life priorities, and standards of labor market success, and impose these perspectives on their young clients. These ideological positions affect the way services are delivered. The Lanier area's label as a "dysfunctional" neighborhood marks many young residents as unfit for career success. Unable to live up to institutional expectations, young people are denied equal treatment within the institutional setting. In Palm Heights, on the other hand, fewer youths are labeled "dysfunctional" and institutional efforts are more effective in assimilating youths to upper-segment identities.

Community-based organizations commonly seek to enhance the labor market prospects of the members of the community they serve. Regarding adults, institutional efforts usually focus on job placement, counseling, and training. For youths, these efforts also involve cultivating attitudes toward education and work and shaping cultural norms of behavior and life priorities (Fernandez and Harris 1992; Harrison and Weiss 1998). Many programs target distinct populations, age categories, and "at-risk" groups in selected geographical areas, such as neighborhoods or school districts. Ethnic minorities, suffering a long history of institutionalized discrimination, often mistrust governmental and nonprofit institutions, and rely more heavily on religious institutions and personal networks (Kasinitz and Rosenberg 1996). Once youths participate in the programs offered in the Lanier area and in Palm Heights, a filtering process has already selected those youths who are more likely to accept public authority and who are more responsive to the strategies pursued by these institutions. Thus, the interaction between the staff of local institutions and their clients is already streamlined in such a manner that resistance on the side of the youths is reduced, giving institutions greater leeway in shaping the work identities of young people.

Rather than examining this selection process, I focus on processes taking place within local institutions. I am especially interested in institutional attitudes that label some youths as ill prepared for educational and labor market achievement. Inner-city school districts sometimes use cultural labels as a justification for the dismal academic achievement rates among minority students. While some students are treated unfavorably based on their cultural identities, others are sorted out as "troublemakers" and subjected to disciplinary procedures if they do not conform to the behavioral expectations of educators (Bowditch 1993; Browning 1994). Although I focus on community-based institutions, not on the school system, I expect that institutional administrators and staff have similar cultural preoccupations and ideological agendas as some educators in inner-city school systems.

Like schools, community-based institutions provide a setting in which youth interact with peers and adults. Institutions influence the formation of peer groups, for instance, by servicing certain geographical areas or by recruiting young people based on perceptions that they are "at risk," or able to "make it."[1] Once youths are recruited into programs, institutional staff often assess and categorize the cultural potential of their young clients. My observations indicate that institutions use their cultural judgment to decide on the choice of services youths receive and the method of service delivery. Institutional practices routinely classify young people into those who have the potential to succeed and those who do not have this potential. Through this categorization process, institutions enable some youths to move into upper-segment careers, while denying this possibility to others. Thus, community-based institutions are not innocent and apolitical service deliverers, but their practices actively contribute to the marginalization of many of the youths they pledge to serve.

Socially produced norms of what constitutes the cultural "mainstream" and cultural "dysfunction" shape the attitudes of employers toward workers. Stereotypical perceptions of cultural requirement for certain jobs lead many employers to rely on carefully constructed ethnic and social recruitment networks to ensure that the workforce has a particular set of values and attitudes toward work (Kirschenman and Neckerman 1991; Mattingly 1999). Other employers exclude ethnic minority groups from applying for jobs by advertising in newspapers that do not circulate in inner-city areas or among ethnic minorities (Turner 1997; Wilson 1996, 111-146). In a similar way, institutional administrators make cultural judgments about their clients. They view lifestyles, attitudes, and norms of behavior as independent forces that constrain or enable career achievement. They fail to situate these characteristics in a greater social context in which lifestyles, attitudes, and behavioral norms signify cultural position but do not necessarily predict labor market competence. For example, grooming or wearing tattoos do not reflect a person's ability to learn or complete certain work tasks (Fernández Kelly 1994), yet these characteristics evoke images of this person's interpersonal skills, attitudes, work competence, and even intelligence.

Embracing the idea of neighborhood effects, many institutional administrators perceive the neighborhood as a spatial incubator in which cultural deficiency is bred. Stereotypes of dysfunction apply to entire residential areas and label the youths who live in these areas as unlikely to achieve in the labor market. If these cultural perceptions influence the scope and nature of services delivered to youth, then institutional practices partially explain why so many youths in stigmatized inner-city neighborhoods are caught in a vicious cycle of social and economic marginality.

The Lanier area and Palm Heights provide interesting case studies to compare the effects of institutional practices on youths. The comparatively dense institutional network of the Lanier area offers a greater variety of educational and employment-related services than Palm Heights. This imbalance undermines claims that marginalized inner-city neighborhoods are simply underserviced. Rather, other circumstances must explain the lower labor market prospects of Lanier area youths relative to Palm Heights youths.

To investigate an alternative explanation, I focus on institutional approaches toward service delivery in the two study areas. I examine, in particular, how cultural representations of the neighborhood affect the practices of service delivery and thereby shape the work identities of youths. This study does more than simply measure the success rate of job training and placement programs; it concentrates on institutional agendas of establishing cultural norms and on cultivating ideological perspectives of work.

Lanier Area Institutions

Administrators of institutions have ideological preconceptions about the Lanier area and the young people growing up in this neighborhood. The interviews revealed that many administrators believe that the cultural context of the neighborhood shapes the career paths of youths. As if they were familiar with the academic literature on neighborhood effects, the administrators suggested that peer networks and adult role models affect young people in the Lanier area negatively. However, they do not perceive the situation as "hopeless." They are convinced that neighborhood influences do not affect all youth in the same manner, but that some youths can escape the downward spiral of cultural pathology that rages in the neighborhood. To make their efforts worthwhile, many institutions categorize youths into achievers—those who may be able to escape neighborhood pressures—and underachievers—those who will inevitably succumb to neighborhood influences. The director of a community center explains:

> You know, there's different extremes [in the Lanier area]. There is the kids: "yeah, I wanna finish high school, I wanna go to college and I

"You're Good with Your Hands, Why Don't You Become an Auto Mechanic?" 59

wanna get out of here." There is those extremes. Then there's people, you know, who love the neighborhood, wanna get a job, wanna stay here. And there is patterns of a lot of unskilled labor. That's very popular in the employment thing, you know, people that do a lot of construction. They're not apprentices, you know, carpenters or anything like that, they just go help their uncle . . . And there's things like house cleaning. That's a popular occupation among women.

Other institutional respondents repeatedly expressed the same perception that there are two categories of youths. The "achiever" group is believed to have certain cultural characteristics that enable youths to succeed in the job market. Usually, these characteristics reflect the cultural norms of the administrators themselves. The "underachievers," on the other hand, are considered culturally "deficient" and "dysfunctional." This cultural selection process is usually not formalized (as, say, program eligibility requirements) but operates on an informal, often subconscious level through cultural judgment on the side of administrators. The administrators' assessment of cultural competence for career achievement is a sorting exercise to determine who should receive services, and what kind of services they should obtain. The problem with this practice is that institutional administrators and staff use ideology-driven judgments to separate the worthy from the hopeless.

Institutions in the Lanier area offer services that fall into three categories: (1) removing barriers, (2) sheltering youths from negative influences, and (3) establishing standards of what constitutes "normal" behavior and labor market success. The first category of services, removing barriers, targets mainly the "achiever" group, consisting of young people whose labor market perspective already meets the standards of institutional staff. These youths tend to receive ad hoc services that remove immediate barriers to career development, such as day care for young mothers or transportation services to work and job-training programs.

Secondly, institutions pursue strategies that shelter both the "achiever" and "underachiever" groups from negative neighborhood influences. Some administrators refer to this strategy as "intervention." Most respondents agreed that intervention is an important strategy that removes mid- and long-term barriers to labor market success. The director of a youth program states:

> I think that intervention is a big key. I think that you have a few individuals that rise above, that come out of the hood, or whatever, and are very successful. But you still have the majority of kids that don't.

The remainder of this interview revealed that the director assumes that all youths start out with the same potential for labor market achievment, but that this potential is continuously corrupted through exposure to the dysfunctional context of the neighborhood. "Intervention," in this case, means protecting young people from neighborhood influences that prevent them from reaching their full potential. The director of another local youth center shares the perspective that intervention is an important strategy of sheltering youths from peer pressure and the influence of local adults. She states that one of the center's primary objectives is "to provide the kids with a safe zone" that protects them from "gangs, people drinking, and fights." This idea of a "safe zone" illustrates the institutional intention of providing a setting that offsets negative neighborhood effects.

The services falling into the third category aim to establish standards for behavior and labor market success; they cater to the "underachiever" group. These services focus intensely on counseling, tutoring, and mentoring. It is within this third category of services that cultural processes of exclusion are institutionalized. Many administrators consider certain behavioral traits as "dysfunctional" and "dislocated," and believe that these behaviors are incompatible with economic or social achievement. Administrators repeatedly cited "living for the day" and "not looking ahead into the future" as dysfunctional cultural traits. The administrator of a training institution explains:

> We have a tendency—we, speaking of this culture—of trying to live for today without even thinking, you know, where am I going to be five years from now. They [Lanier area youth] need to make it through this week, they survive this week. They don't really think about the following week. You know, it's like they live from day to day. And so they're looking at maybe, hopefully graduating from high school but a lot of them have no idea what they wanna do after high school. That's too far in the future for them.[2]

An education provider servicing women on welfare expresses a similar cultural inferiority argument, citing the lack of skills to cope with crisis rather than the culture of living for the day:

> We think about our clients living in crisis all the time and they really do for the most part. I think poverty and welfare lend themselves to crisis. They can't maintain jobs because the first crisis that comes along they throw up their hands. And you know, we too have crises in our lives. You know we've gone through all kinds of crises as employees but we learned some way to be able to handle a crisis because crisis doesn't keep us off track.

The life priorities of Lanier area youths are another cultural trait believed to be incompatible with career development. The executive officer of an educational institution elaborates:

> One of the main things that we see as barrier to education is that even though education is one of their main priorities, it is not number one in their lives. . . . There's certain priorities that a person has: work, take care of your children. Education is maybe number five or six. . . . They come in for three months [to class] and then I don't see them anymore. What happened? "Oh, my child got sick."

These institutional views are remarkably close reflections of Oscar Lewis's culture-of-poverty concept from the 1960s. Living for the day, the inability to cope with crisis and a misguided sequence of life priorities are seen as "dysfunctional" cultural traits that supposedly explain economic and social marginality. However, access to opportunities in the labor market is relatively independent from future expectations, crisis-coping skills, or life priorities. Establishing a family as a single teenager, for instance, is only incompatible with continuing education when appropriate and affordable services for single parents are absent. Without accessible day care, many young parents are forced to discontinue education. A culture-of-poverty perspective lays the blame for failure on having children. However, if a young person has access to day care, this person would have a much greater chance to attend college and to prepare for a professional career despite a lack of future perspective, crisis coping skills, or life priorities that favor families over careers.

Many institutions attempt to communicate cultural standards to their young clients. Strategies aim at eradicating "dysfunctional" cultural traits that supposedly damage labor market prospects, and acculturating youths to the norms and values held by institutional administrators. The executive director of a youth center, for instance, explains that she attempts to change her young clients' "mind-set" to match her own life priorities, work ethic, and educational aspirations. She complains, however, that young people do not spend enough time at the youth center under the supervision of her staff to fully offset the negative influences of local peers and neighborhood adults. She says:

> You know, the few hours that we get them here is not enough, unfortunately, to change their whole mind-set, you know, in terms of what they've already seen and been exposed to—and we try. You know, we're all here and we're all working; and you know, we're not living in the projects. And we talk to them about how we went to college and things like that. But the kids I think identify more with their own family

and their own neighbors than they do with us. They don't see us, you know, living next door to them.

An important institutional strategy to shape youths' behavioral patterns is to provide adult role models who reflect cultural standards deemed appropriate. For instance, one organization invites guest speakers who stress the importance of having a "vision to be someone." The meaning of "being someone" is synonymous with acquiring social and economic status, measured by the yardstick of the cultural norms embraced by administrators, but not necessarily by the Lanier area community. Most institutions, however, do not feature outside guest speakers with regular frequency. Rather, institutional staff assume this role-model function themselves. In this function administrators shape the aspirations of young people and manage the cultural assimilation to the external values and norms held by the administrators. A career counselor responds to the question of whether he and his colleagues consider themselves to be role models for their adolescent clients:

> Yes, we do a lot of times because you have a kid that you [supervise] and he has no aspirations of going to college or thinking that he can make it. . . . Now, here you have a kid that is looking up to you and says: "Well, maybe it is true, maybe I can make it in college."

Lanier area institutions are sites of cultural manipulation that prepare young people for social and economic upward mobility through promoting a set of external values and norms—but they also exclude nonconformist youths as unfit for social and economic achievement.

But what are the cultural norms that institutions promote? Institutions embrace particular interpretations of what constitutes labor market success, and these interpretations are not necessarily shared by Lanier area families and youths. Two youth program supervisors, for example, consider a restaurant job and the lack of college education a career failure. One of the supervisors says about a youth in their group:

> Peter [not his real name] graduated from high school. His mom is the greatest mom in the world, she is the sweetest mom. But she doesn't push him, she doesn't push him. "Peter, right now, is working at Bill Miller [a restaurant chain]. If that's what he's happy with, I'm happy with him." She should have tried to push him. That would be the difference.

Her coworker continues:

> Instead of getting a [college] scholarship for [athletics] he decided "I rather make those 100 bucks a week and get myself some jeans, nice shirt, and have some money to go out during the weekend." Things that

kid has never had before. And his mom: "if that's what [he] wants then I'm happy."

The course of the interview revealed that the supervisors are unaware that among local families the income of a youth often provides an important contribution to the family income. In addition, making a contribution to the family income marks the transition from adolescence to adulthood and earns them status within the family, the community, and among their peers (Fernández Kelly 1994). In a local context, choosing work over a college education is not necessarily seen as a failure but as assuming adult responsibilities. Many institutional administrators, however, do not share this perspective.

The cultural norms and values held by administrators affect the manner in which services are delivered to youths. For example, a role play in a youth counseling program aimed to teach youths how to encounter the stereotype of being a predatory gang member. Two staff persons assumed the play roles of a storeowner in the mall and a male youth, who is a customer in the store. The staff person who played the youth pulled down his pants several inches to suggest that he is wearing oversize jeans, turned his baseball hat sideways, let his arms and shoulders casually hang forward, and began mumbling in typical West Side slang. The play communicated to the youth group that their styles of dress, behavior, and speech do not reflect proper codes of conduct, but signify that they are unemployed school dropouts and gang members with no good intentions. The youths are being told that most people would identify them as shoplifters. The message conveyed to youth is that they can only achieve social and economic status by changing their styles of dress, behavior, and speech.

Many administrators believe that the origin of "dysfunctional" attitudes and behavior among youths is the neighborhood environment. Some go as far as rendering the entire Lanier area as "dysfunctional." With this spatial perspective of "dysfunction" these administrators believe that most young people from the Lanier area are, from the outset, doomed to serve in the secondary labor market simply because they live in the Lanier area. One youth group organizer is convinced that local youths do not qualify for jobs in the upper segments of the labor market. He states:

> Why I am upset is because to me a pediatrician from that neighborhood wouldn't be realistic. A doctor to me is unrealistic [. . .] Well, one of them wants to be a nurse and I can see that as being realistic. Especially

that program Project Quest[3] gears to disadvantaged neighborhoods. They're nutritionists, plumbers; they're all associate degrees. We're not talking about PhDs or doctors.

A community activist criticizes local institutions exactly for promoting these kind of myths about the labor market limits for local youths based on neighborhood stereotypes:

> Yeah, I think that in the last 30 years, a lot of the guidance counselors would send them [Lanier area youths] into trade, trade skill occupations or direct them in that way. "Oh, you're good with your hands, why don't you become an auto mechanic"—or something like that. I think that they're directed that way. That's not necessarily the way they all want to go.

The efforts of community-based institutions to assist career development now become counterproductive. In response to the perceived lack of cultural capital in the neighborhood, institutional practices channel youth into secondary occupations and deny them access to careers in the primary sector of the labor market. Ironically, the occupations that administrators consider "good" for local youths are occupations that are considered inferior according to the very norms the institutions hope the establish.

Institutions provide a setting in which internally constructed work identities collide with external representations. The difficult choice for institutions is to promote either external standards of work, and thereby deny young people's internal cultural identities, or to support internal work identities, and thereby channel young people into secondary occupations. Institutions cannot easily resolve this conundrum rooted in the place-contingent nature of standards of work and ideas of success and failure. In the name of pragmatism and the advancement of individual clients, many Lanier area institutions opt for encouraging their young people to adopt norms and values of the mainstream.

But institutions do not blindly reiterate the norms and values that are emphasized in popular discourse. Contrary to my expectations, for example, the majority of institutions are unwilling to reject young pregnant women and young mothers on the grounds that they are culturally dysfunctional. This finding surprised me because the academic literature stresses the correlation between teenage and out-of-wedlock childbirth and labor market underachievement. Lanier area institutional administrators apparently do not share the viewpoint of these researchers. They do not perceive teenage and out-of-wedlock parenthood as a cultural attribute that causes career failure. Nor do they respond with corrective programs meant to change the sexual behavior of young women and men. Instead, they would like to see more support services for young parents, such as day care; but they lament that resources are too sparse to provide these services in an adequate manner.[4] This particular finding challenges the idea that institutions simply embrace the norms and values that conservative academic and public discourse presents as "mainstream." Rather, institutional perceptions of norm behavior and labor market achievement reflect evolving ideological worldviews that draw on local as well as wider social discourse and on the particular ideological inclinations of administrators.

Palm Heights Institutions

Many institutional goals, strategies, and practices in Palm Heights are similar to those in the Lanier area. For example, most institutional administrators adhere to similar norms of what constitutes labor market success and similar beliefs of how success can be achieved. However, an important difference is that few youths in Palm Heights are perceived as culturally "dysfunctional." Youths are therefore not classified into opposing groups of "achievers" and "underachievers." A youth counselor points out that no group of youths is singled out for the provision of special services. Rather, the organization provides support to all youths in the group. The counselor says:

> We want them [all] to be productive. And like we're saying: "maybe college isn't for you." But we're not going to say: "OK, nobody has to go to school anymore." It's mainly helping them think about what they want to do.

An administrator of another community center believes that most young people have similar chances in the labor market. In her eyes, age, rather than attitude and other characteristics, determines whether a youth gains access to certain kinds of jobs:

> A lot of kids have dropped out of school. It's an individual thing. It depends on the situation, you know, rough times at home. They go through different stages, they end up dropping out [of school] and working. Well, I think McDonald's is the job for a young kid anyway. They may want more but they can't get more because [of their age].

As a result of this focus on inclusiveness, one group of young people is not favored over another, but rather all youths are considered equally fit for the labor market. All program participants are encouraged to pursue their individual career goals.

Nevertheless, the inclusiveness of Palm Heights institutional practices has its limits. For instance, Palm Heights institutions struggle to retain young women once they are involved in intimate personal relationships. A local youth group administrator explains:

> The girls are hard to keep [in the youth group] because they get a boyfriend and it's like boom their gone. And that's not very good for us because, you know, we've lost some really good girls because they get a boyfriend. And as soon as they dump their boyfriend they're back again.

The tendency among young women with boyfriends to opt out of the program creates a self-selection process, which indicates that institutions do not respond adequately to the needs and interests of all youths in the neighborhood—in particular those young women who wish to focus on issues of emotional attachment, intimate personal relationships and family.

Similar to the Lanier area, institutional initiatives in Palm Heights attempt to shelter young people from neighborhood effects and establish norms of behavior and standards of career success. In both areas, for instance, community-based institutions attempt to control neighborhood influences through intervention. An administrator of a youth program in Palm Heights explains:

> Mostly, it [the youth program] focuses on protection from the neighborhood, activities that are going on at school, maybe peer pressure and then intervention for the kids that are involved in that stuff, so that they can get away from it.

In contrast to the Lanier area, however, interventionist practices by Palm Heights institutions target youths of a certain age category rather than a group of youths selected based on their career potential. A supervisor of another youth group explains:

> With the older kids we try to [target labor market objectives] a lot. But with the younger kids, the focus is just on staying out of trouble. So, it's two entirely different things. But with the older kids, yeah, we're very much focusing on getting ahead and thinking about school.

Palm Heights institutions stress the goal of relative improvement rather achieving a set level of academic excellence that is defined in absolute terms. This goal of relative improvement can be applied to all youth at different academic levels. An administrator points out:

> We've taken kids that are pretty high risk and they weren't kids that were doing good in school but it was an opportunity for them to see

what would happen if they try to be better in school. So it's not selective as far as grade point average. That's one thing that we're really different. I know that some other groups have all grade point averages and we just disagree with that entirely because we feel that, you know, we're here to help anyone who wants to be here. A lot of the kids are here because they want to and not because it's part of their GED.

Despite the differences in approach, both Lanier area and Palm Heights institutions attempt to establish standards of cultural and behavioral norms. Palm Heights administrators assume that a youth's chances for labor market success improves greatly if he or she adopts the behavioral norms, values, and attitudes toward work which administrators attempt to instill. Administrators indicated that they are rather successful in engraving cultural norms into the behavioral patterns of their clients. A youth group counselor proudly describes the labor market achievements of her former clients, who successfully completed the program and learned important norms and values:

You know what so interesting is? A lot of them are going to college. And it's so funny because some of them would be counselors now. So, it's real interesting. Some of the boys wanna go into sports medicine, physical therapy. A couple of these kids are talking about going to the Air Force.

Some administrators credit their efforts of shielding young people from the negative adult role models in the neighborhood for their success in making advancement into upper labor market segments a real possibility for their young clients. Organizing youth groups by age isolates participants from the influence of negative adult role models and older youths. This strategy is effective in controlling neighborhood effects. One administrator observes:

They [youths] are mostly in their age group. I don't see them anymore, like these younger kids, hanging out with older people. They used to but not anymore. What, about 15 years ago, you had older guys here and the younger kids would follow them. Not anymore!

The institutional perspective on cultural processes within the neighborhood is that local adults and older adolescents are culturally unfit to be appropriate role models for youths. Therefore, Palm Heights institutions replace neighborhood peer networks and adult role models with peers and adults from other geographical areas. They make an extended effort to break the spatial isolation of young people and create an institutional environment in which youths interact with adolescents and adults from other neighborhoods. This spatial outreach exposes young people to cultural contexts beyond their own neighborhood. A youth group administrator describes the outcome of this spatial outreach strategy:

> We're trying to get all of the South Side. Burbank, Harlingdale, McCollum, South End although we don't want to reach into other church parishes. We don't want to go into the South End cause that's another parish. But we got kids that are from MacArthur, which is like way on the North Side. We've had kids coming from everywhere in this town.

A Palm Heights community center mixes its athletic teams with local and nonlocal members and reach out to other community centers to bring young people from different neighborhoods together. The director of the center remarks:

> Mostly the older guys are from another part of town, the East Side, the West Side. They drive here. They come over here cause they like the players. We compete in a league here, or have other teams come to our league. We have teams from Normoyl and Quale [community centers located outside Palm Heights].

This strategy of geographical outreach goes hand in hand with a strategy of introducing people to the program who represent the cultural mainstream. An administrator proudly draws attention to the nonlocal adult volunteers she was able to recruit for the program and the degree of exposure youths receive to these adults:

> They [volunteers] are all from different backgrounds. One of the ladies that helps is a juvenile probation officer, and another lady that helps is a retired agent, another guy works for the FBI, another lady that helps works here at the directory, and then we have another student who helps—I think he's around 20—and he comes around and helps. And then there's a guy from Boston here helping us. Actually, we spend a lot of time with those kids. I mean, quite a bit. We see them at least, I would say, four times a week. And that's a lot when you think about it.

Furthermore, Palm Heights administrators emphasize their success in recruiting young people from different cultural backgrounds. For instance, one program administrator sees it as a particular accomplishment that her group includes youths from both private Catholic schools and public schools. She explains:

> We've had a lot of luck in getting kids from the Catholic schools. Which is real interesting because usually it's real hard to get Catholic school and public school kids together because they think they're real

different from each other and we've been real lucky in getting Catholic school kids. We can blend them and that's pretty good.

The respondent feels that this "blending" exposes group participants to alternative cultural perspectives, affecting their labor market objectives and motivations in a positive way.

Palm Heights administrators believe that the strategy of recruitment of youths and adult volunteers from a wider geographical area and from "mainstream" cultural backgrounds protects their youth groups from negative neighborhood effects. Through this strategy, administrators suggest, young people are introduced to a wider set of norms and values not available within the immediate neighborhood. Relying on an extended network of people and a citywide cultural reference system, Palm Heights institutions appear to be less threatened by the influence of neighborhood context. Administrators are more optimistic that all youths can escape the imperative of neighborhood effects, and they are reluctant to reject young people who do not embrace "mainstream" perspectives of behavioral norms, values, and career achievement. An institutional administrator explains that her group's goal is "mainly helping them [youths] think about what they want to do [for a career]. It's not only about the future but also to mold them culturally." The strategy of inclusion and spatial outreach diminishes cultural isolation and encourages assimilation to external cultural norms. But Palm Heights institutions also demand that youths leave behind the cultural identity of their neighborhood in order to achieve what society deems as success. Local and "mainstream" identities appear to be mutually exclusive.

Conclusion

Young people who adopt "mainstream" cultural traits and worldviews are, indeed, more likely to remain in school and attain professional careers than youths who maintain subcultural identities. However, increased labor market prospects are not associated with learning superior behavior in an institutional setting where people who live in different neighborhoods and come from different ethnic and class backgrounds interact with each other. Rather, institutions merely help young people to adopt norms, values, and behavioral patterns that are subject to diminished forces of cultural discrimination in the school system, the labor market, society at large, and in the institutions themselves. Career chances do not improve because youths overcome cultural inferiority, but because they adapt to external expectations.

In both neighborhoods institutional administrators harbor essentialist conceptions of culture. Cultural dysfunction, they believe, renders young people incapable of labor market success and causes failure. Culture is not acknowledged as a force of social distinction that marginalizes those who are different. Rather, culture serves as an explanation that legitimizes the exclusion of young people who do not conform to dominant norms. Institutional practices subsequently erect cultural barriers to the labor market that are difficult for young people to overcome. Since these practices respond to the demographic composition and the behavioral patterns of residents, they constitute a particular form of neighborhood effects.

Lanier area institutions confront a dilemma because they apply external standards of behavior and economic achievement to local contexts that embrace different cultural norms. If they support local work identities they push young inner-city residents into economic marginality. If they promote upper-segment work identities, they request local youths to acculturate. This dilemma relates to the popular perception that the economy is a "natural" process, whereby the invisible hand of the market compensates workers for their labor. This perspective does not acknowledge the discursive practices revolving around cultural struggles for domination which shape labor reward structures, and which devalue the labor of inner-city minorities.

Palm Heights institutions also perceive a great threat emanating from neighborhood context. Although they embrace similar cultural interpretations of what constitutes success and how it is achieved as Lanier area institutions, they use different approaches in dealing with the culturally "deprived." While Lanier area institutions exclude these youths from institutional nurturing, Palm Heights institutions focus on spatial outreach and social inclusion and provide "bridges to other social networks" (Fernández Kelly 1994, 109). But this strategy of inclusion and assimilation is highly problematic, because it feeds into a wider agenda of cultural colonization. In both neighborhoods internal representations of work identity are subordinated to the external perspectives of work and life priorities. Young people are asked to deny the internal cultural identities of their local community and adapt the external identities that exist on a wider spatial scale.

The issue of spatial scale occupies a central position in the manufacturing of cultural difference and in fostering assimilation. A youth counselor in Palm Heights explains that she wants to move her clients out of the sociospatial reference of the neighborhood into a wider urban reference scale. She states:

It is helpful for them [youths] to see that there is a different way of doing things and that there are more role models [outside the neighborhood]. It keeps them tied into what's happening in our city . . . quite frankly that's the community that's so important for [our] youth in terms of the welfare of the youth and the future of the youth.

This selection of scale of reference is a major difference between the institutional practices in the Lanier area and Palm Heights. Lanier area institutions are entangled in conflicting cultural norms that exist on the neighborhood scale and a wider geographical scale, while Palm Heights institutions have "jumped" scale and embraced norms, values, and identities on the citywide scale. Chapter 5 examines the issue of scale in relation to the neighborhood as a spatial unit that expresses and produces cultural difference.

Notes

1. Furstenberg et al's (1999) *Managing to Make It* indicates that this vague terminology is commonly used in reference to the social mobility of inner-city youths. The terms avoid a precise definition of success and failure, and the circumstances that would lead to social mobility. Similarly, when administrators of community-based institutions use the term "at risk" they did not rely on consistent or well-thought-through definitions for this term.

2. Note that the respondent switches from "we," identifying with local culture, to "they," distancing herself from the attitudes of local youths. Her schizophrenic speech suggests that her identification with the local culture is at odds with her role as an institutional administrator.

3. Project Quest (Quality Employment through Skills Training) is a local job placement and training program.

4. Edin and Lein (1997) have noted the well-developed network of private charities and churches willing to support single mothers in San Antonio.

Chapter 5

Between Scales: Agency and Ideology

Ideologies of Place and Scale

The neighborhood is an idea, which we use to understand socioeconomic relationships in the city. We deploy this idea to frame our own identities and those of others. Public and academic debate, for example, has treated the neighborhood as an essentialist spatial unit, an ontological actuality, that separates marginality and dysfunction from the rest of society. This separation supposedly occurs in a natural manner, like when the oil separates from the vinegar in our salad bowls. While in the previous chapters I also used the idea of the neighborhood to frame internal and external identities, I now problematize this spatial category. The neighborhood is not an ideology-free spatial category, but rather a discursive construct that is created precisely for the purpose of representation and classification. In this chapter, I acknowledge that cultural identity is not ontologically fixed to neighborhood context. In particular, I turn a critical eye on the scale of the neighborhood and the coupling of the idea of the neighborhood with cultural dysfunction.[1]

The geographical scale of the neighborhood enables the presentation of cultural dysfunction as a problem that is spatially contained and existent only outside the core space of society. Mainstream space is defined in opposition to the dysfunctional neighborhood and exists on a wider scale. In public and academic discourse the scale of the mainstream is usually not specified because the cultural norm is not imagined to be restricted to a particular spatial context.[2] For analytical purposes, however, alternative scales that express cultural identity can be defined by the city of San Antonio, the state of Texas, or the country of the United States. In the following analysis, I focus on the Lanier area as a neighborhood-scale representation of dysfunctional space, and contrast this scale with the urban scale of San Antonio, representing the cultural mainstream. I admit that I lacked a clear guidance for selecting an opposing scale to the neighborhood because the "mainstream" is an unmarked and aspatial category. However, people tend to identify with the cities in which they live, and in most cities residents have an image of the cultural characteristics they share with their fellow inhabitants. I therefore assume that there is a degree of continuity regarding cultural identity within the urban context of San Antonio, and that the external identity of the Lanier areas reflects a perspective from this urban scale.

To gain a better understanding of the linkage between the concepts of identity and scale, I also examine the relationship of both concepts to ideology. I define ideology as a particular interpretation of the social world that supplies a blueprint of general relationships that explain the working of society. The cultural resources that constitute ideology are spatially referenced and embedded in concrete geographical contexts (Gregory 1978; Thompson 1984). Pinderhughes (1997) examines neighborhood youth groups in New York City and observes that "community-based ideologies supply information to youth about how to interpret their experiences, their identities, and their relationships with other groups" (17). These youths perceive community-based ideologies as fixed societal norms, and deploy these ideologies to articulate their own cultural identities and those of others. The significance of Pinderhughes' work lies in his treatment of identity, ideology, and community as an interlocking system. Since this system is highly contingent on the idea and material context of the neighborhood, I refer to "neighborhood-based ideologies," which produce labor market identities and norms of achievement.

But neither communities nor neighborhoods are independent worlds of their own. Throughout the course of their daily lives, individuals maneuver different spatial contexts. One particular way in which different ideologies are brought in contact with each other is through overlaying spatial scales, which frame different "conceptions of reality" (Delaney and Leitner 1997, 94-95). My point is that locally situated individuals or social groups confront multiple ideological frameworks that exist on different spatial scales. A youth growing up in the Lanier area thus engages with both neighborhood-based and citywide ideological frameworks.

Focusing on the neighborhood and the urban scales, I examine how the identity of Lanier area youths is linked to scales of representation and how young people position themselves in relation to opposing scales of representation. Of particular concern is the ability of young people to "switch" scales in order to influence their own labor market outcomes.

Between Scales: Agency and Ideology 75

Much of the existing literature on scale focuses on how economic and political struggles are being dragged onto another scale to mobilize resources and to strategically outmaneuver opponents (Harvey 2000; Herod 1991; Miller 2000; Smith 1984). But we know little about the processes through which young people articulate their identities in reference to the neighborhood and other spatial scales, and how they strategically select these scales to achieve particular aims (Aitken 2000).

This discussion centers around the role of human agency, or the "capacity to 'make a difference'" (Giddens 1984, 14). Although recent poststructural theory has contested the idea that agency can be isolated from discourse and analyzed separately (e.g., Pile and Thrift 1995), I think that maintaining agency as an analytical category is important for envisioning political strategies of resistance and empowerment. Only as agents can inner-city youths, community activists, and urban institutions intervene in the social processes that affect their lives.

A pressing question is how Lanier area youths resolve their exposure to opposing ideological frameworks that exist on the neighborhood and the city scale. Can they articulate their identity in relation to one or the other scale, and how does this affect their situation in the labor market? In other words, can agents strategically jump scales of representation as a means to shape their own labor market destiny? Figure 5.1 visualizes how agents choose between scales to position themselves in different ideological frameworks that underpin identity formation.

The Lanier area is particularly suited for this analysis because internal work identities differ from external expectations. This situation enables me to juxtapose ideologies on the neighborhood and urban scales against each other. In contrast, young people in Palm Heights have largely internalized external ideology and assimilated to external labor market expectations. If work identities differ little between Palm Heights and its surrounding residential area, it would be difficult to observe scale effects. The analysis that follows therefore focuses on the Lanier area.

Figure 5.1: Ideology, Identity, Agency in Spatial Context

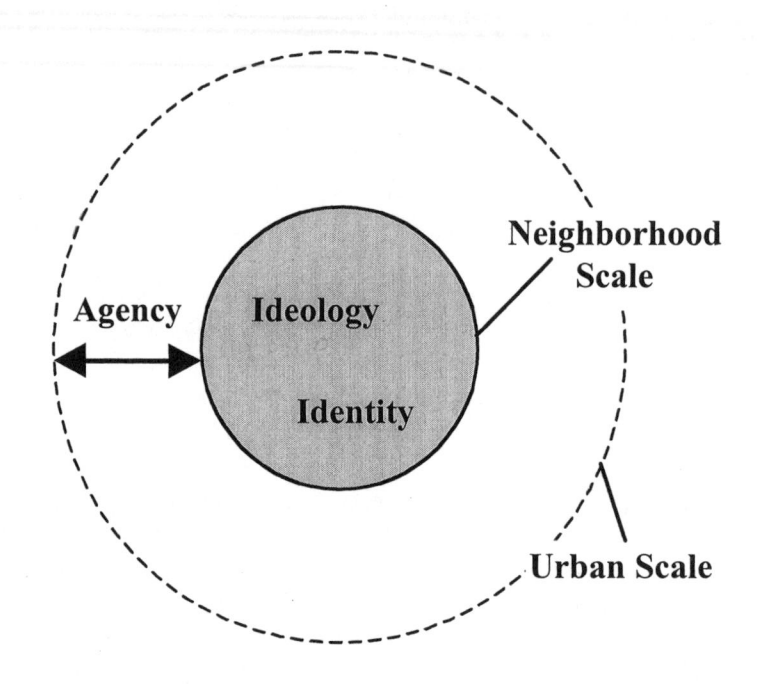

Agency ↔ Ideology / Identity
Neighborhood Scale
Urban Scale

Between the Barrio and the City

Lanier area youths embrace what Pinderhughes (1997) calls community-based ideology for interpreting the social and economic circumstances surrounding them. The neighborhood provides the spatial reference for this community-based ideology. Lanier area ideology is shaped in response to the material context of the neighborhood. For example, conceptions of work and career are formed on the basis of the jobs that are available within the neighborhood. The youth program director explains why conceptions of labor market success focus on secondary occupations rather than professional jobs:

> Because their world is so small it [their career objective] is gonna be determined by what's happening in the community and who they see.... And if you drive around, what you see are tire-shops, are taco-restaurants, are, you know, stuff like that. But if you were just to look around in this area you see that workwise what's available, you know, that's very limited. There isn't very much around. . . . Do you see banks? There are no banks in the area.

The resulting local interpretation of what constitutes a "good job" differs from interpretations that prevail elsewhere in San Antonio where professional and managerial jobs are visible and accessible to young residents.

Gender roles are also embedded in a neighborhood-based ideological framework. Lanier area gender roles differ from those that exist in other parts of San Antonio. The administrator of an education and training program observes that the Lanier area's common conception of motherhood is shaped by an ideology that is rooted in "Mexican" tradition, and this tradition suggests that women carry the main household and child-rearing responsibilities. Whereas other Latino communities in San Antonio have adopted a liberal ideology of gender equality in education and the labor market, the Lanier area maintains a traditional ideological conception of the social role of women. The administrator refers specifically to the Lanier area when she states:

> I've seen some instance, when a young girl becomes pregnant, living at home: "You have to stay home, you're a mother now. You're not gonna go to school. You're a girl anyway. So therefore you're not going out. You're gonna marry some guy and he's going to take care of you." I think some instances are like that; especially among the Mexican Americans [in the Lanier area].

In a similar vein, a former Lanier area resident recalls that her local role as a Latina woman would have required her to quit school and "work around the house." When she moved to another neighborhood on the West Side, a different ideological frame of reference defined a new gender role for her. There she was no longer expected to become a housewife, and instead was motivated to attend college and focus on the development of her career. Today she owns and manages a nursing home, a career which would have been inconceivable had she stayed in the Lanier area. These two examples demonstrate that the meaning of the terms "mother" and "Latina woman" change in different spatial contexts of reference. Meanings differ even between West Side neighborhoods.

While the Lanier area frames ideology on the neighborhood scale, the context of San Antonio projects a different ideology onto the urban scale. Different scales of reference can facilitate the formation of distinct cultural identities for young residents of the Lanier area. For example, the ethnicity of Lanier area youths assumes different meanings on the neighborhood and the urban scales. The Lanier area is more than 95 percent "Hispanic." In this ethnically homogenous area, neighborhood-based ideologies do not articulate strong conceptions of ethnic and racial difference. The school district, representing the urban scale, however,

Chapter Five

draws a much larger variety of students, teachers, and staff with various ethnic backgrounds. In school the ethnicity and race of a young person become important markers of cultural identity. An eighteen-year-old college student explains that, within the Lanier area, she and her local peers never think of themselves as "ethnic." But in school, where ethnic and racial differences exist, she and her peers are labeled "Hispanics." The ethnic identity of young people and the emphasis on ethnicity changes with the spatial scale of reference.

The concepts of ethnicity and gender operate in tandem to define the cultural identities of young people. Within the context of the Lanier area, conceptions of the family and the household situate women in the roles of caretakers, housewives, and secondary workers. Once the cultural context of reference is expanded beyond the neighborhood, different ideologies not only redefine the identities of these women, but many women are no longer prepared to maneuver the cultural context of this wider scale. The relative isolation of Lanier area women (resulting from the very expectation of "working around the house") has kept these women from learning the cultural codes that are necessary to function in social context outside the Lanier area. My conversation with an eighteen-year-old housing court resident revealed how gender and ethnic identities interlock to confine Lanier area women to the neighborhood scale, and deny exposure to external cultural conventions and norms:

Resident: My friend Wayne [not his real name] never really knew his dad and his mom's boyfriend was to him his father. But the parents didn't speak much English; and all his brothers they all spoke English but his sister had this fair English.

H.B.: Why the sister?

Resident: I think one of the guys was in the military, the oldest one, and when he came back he wanted to make sure that all his brothers were, you know, grammatically correct because he knew how things were.

H.B.: But not the sister?

Resident: She was like, from tradition she was not real educated or whatever, so he was more concerned with his brothers than he was with his sister.

The neighborhood-particular gender role confined the sister to the spatial context of the Lanier area. The sister's lack of standard English skills is not problematic as long as she stays within the Lanier area where she can communicate in Spanish or a strong "West Side dialect."

If, however, she visits other parts of San Antonio, her lack of standard English mark her as "illegal Mexican." In contrast, Lanier area men, in their role as breadwinners, have wider activity spaces and are expected to interact with people outside neighborhood boundaries. Proper language skills as well as knowledge of dominant cultural codes are essential to maneuver this scale effectively.

On a wider scale, conventions differ from those in the Lanier area. According to the prevailing worldview in the Lanier area, the role of women emphasizes the family, household, and child rearing. Wider-scale ideology, on the other hand, promotes gender roles that are more independent from the family and include higher expectations from the labor market. Embracing this liberal worldview, the Teen Unit of the Texas Workforce Commission actively persuades teenage mothers to stay in school and focus on the prevention of pregnancies. This agenda reflects an ideology placing greater value on work and labor market advancement than on being a mother and child rearer.

These wider-scale ideologies not only establish positive norms of work and career, they also create negative conceptions of cultural dysfunction. Well-known examples of demonized behavior include teenage pregnancy, out-of-wedlock childbirth, welfare dependency, living in public housing projects, and dropping out of school. An institutional infrastructure is in place in San Antonio to systematically isolate individuals and groups who exhibit these traits and mark them as culturally dysfunctional. The San Antonio police force, for example, participate in enforcing the marginalization of culturally different groups. I accompanied a police officer on his eight-hour shift patrolling the inner city of San Antonio. The officer spent a large portion of his time responding to calls from property owners to investigate "suspicious-looking" people for trespassing or loitering. What made these persons suspicious looking was that they were not at work in a weekday afternoon, their Latino ethnicity, the usage of Spanish, and their bodily styles and behavior that did not adhere to middle-class norms.

Similar to the way these individuals are stereotyped, the spatial context of the Lanier area symbolizes dysfunction. Scale-related processes of differentiation contribute to the negative image of the Lanier area. The Lanier area is large enough in population and geographical area to be visible and identifiable but too small to be a threat as an out-of-control problem. Scale is strategically deployed to construct a space of dysfunction that is contained and manageable. Thus, the spatial representation of marginality on the neighborhood scale legitimizes wider-scale ideology by producing a relatively small and nonthreatening opposite. The stigma of the Lanier area as the dysfunctional "Mexican barrio" enables the positive identification of San Antonio as a whole as the cultural "mainstream."

Wider-scale ideologies use the same markers to label young people in the Lanier area as dysfunctional, which the youths themselves use to construct their self-identities. For example, many young people in the Lanier area express their identity by wearing baggy pants, shaving their heads above the ears and neck, incorporating Spanish terms into a distinct dialect, and performing certain styles, body language, and behavior. By themselves, these markers do not necessarily symbolize dysfunction. Simply wearing baggy pants is not always interpreted as a negative sign of dislocation. If a young person on the North Side wears baggy pants, for instance, this person is still seen as a "normal" kid. However, when these same markers occur in the spatial context of the Lanier area, they become symbols of dysfunction. A seventeen-year-old Lanier area youth explains that his dress style falsely suggests to the police that he is a gang member:

> *Youth*: [The police] stops you for no reasons if you walk around at night. Especially [because of] the way we dress.
>
> *H.B.*: What does dressing have to do with it?
>
> *Youth*: That's how a lot of [gang members] dress. They dress, like they wear their pants all the way down to their knees. But we don't, we got big pants. They're called Chicos.

The spatial contingency of cultural meaning lies at the core of the social and economic exclusion of Lanier area youths. Whether or not behaviors, styles, and traits are interpreted as socially acceptable or dysfunctional depend on the spatial context of reference in which they occur. Scale is a categorization system that references the identities of young people in an ideological context.

Lanier area youths are now situated in two scales of opposing ideological contexts of reference. On the local scale, Lanier-area-based ideologies establish local cultural norms and define conventions of family life and career success. On the urban scale, the ideological of San Antonio imposes a different set of cultural norms, one that renders neighborhood-based cultural conventions dysfunctional. Lanier area youths are caught between these scale-based ideologies. Although Lanier area youths have restricted activity spaces, their spatial confinement is not hermetic and they are exposed to both scales, through, for example, the school system and neighborhood-based institutions. By maneuvering between different scales, they confront ideological contradictions, and they realize that these contradictions are at the root of their marginalization.

Jumping Scale into Assimilation

The youths I interviewed are well aware that different ideological frameworks exist in the Lanier area and the wider scale of San Antonio. Furthermore, they strategically position themselves in relation to these ideologies, depending on their specific situation and personal goals. In matters relating to the peer group, the family, or the local community, they are likely to endorse neighborhood-based ideologies. In situations that regard education or the labor market, they often articulate their identities in relation to an external ideology.

My interview guide contained the question: "If you had any advice for younger kids in the area, what would you tell them?" Eighteen out of twenty-one youths responded with answers relating to education, such as "stay in school," "don't drop out" or "finish high school." These answers reflect a wider-scale ideology that values educational credentials. A seventeen-year-old Lanier senior remarks: "Without education you're nothing. Education is the first key to your success. Without education you won't be able to do anything. You could be a janitor." In the context of the formal interview and in the presence of myself, an outsider to the local community, this particular youth rejects the idea that janitor is a respectable occupation. Yet, this particular respondent dropped out of high school only a few months after he made this statement. In a similar vein, many young women added responses such as "don't become pregnant." Yet, many of these women were mothers. Furthermore, when the interview switched to topics relating to the local community, most of these mothers said that they do not regret having children, and that they were proud of their roles as mothers. This role gives them responsibility and raised their status among peers and within the community. Nevertheless, they realize that this very role compromises their prospects for success in the labor market.

Being simultaneously immersed in different scales of place, young people confront ideological double standards. On the one hand, they embrace locally sanctioned norms of work and family and seek to fulfill locally embedded expectations of work and family. On the other hand, they know quite well that these norms mark them "dysfunctional" on the wider scale of San Antonio. They also realize that the blame for their social and economic marginality is affixed to the neighborhood-based ideology they embrace. To escape the label of cultural pathology, they deploy alternative identities which conform to the dominant ideology on the urban scale.

The jumping of scale is a strategy that young people can use to reposition themselves in the labor market. When youths reject Lanier-area-based ideology in favor of wider-scale cultural norms and values, they are more likely to finish high school, are less prone to become pregnant, and can develop labor market prospects that provide access to upper segment occupations. For instance, the owner and manager of a nursing home, whom I mentioned earlier, left the Lanier area in order to situate herself in a different ideological context, which allowed her to attend college, pursue a professional career, and become an independent entrepreneur. In this case, the decision to relocate was a decision to switch scales of reference in order to gain access to labor market opportunity.

A related strategy to reshape career trajectories is to challenge existing neighborhood-based ideology. This strategy is pursued by a young college student who remained in the Lanier area but resisted outside stereotypes that position her in the secondary labor market. She explains that she was able to advance to college only because she contested the label of the "Hispanic underachiever." Most of her peers were unable to resist the pressure from their families and peer groups and dropped out of school. She also explained that she received most of her encouragement for academic achievement from within the community, especially a Roman Catholic church-based youth group. However, while churches are community-based institutions, they do not necessarily represent community-based ideology. Rather, they endorse norms of life achievement and labor market success that derive from wider-scale ideology. Thus, the interviewee's career aspirations and educational motivation can again be traced to nonlocal ideology. This second strategy, therefore, also represents a jumping of scale, although the person remains physically located in the Lanier area. A third strategy is, of course, for a youth to make a conscious decision to reject wider-scale ideology and embrace Lanier-area-based worldviews. The sacrifice of this strategy is social and economic marginality and isolation from the dominant culture that exists on the urban scale.

Despite the possibility that youths can actively influence their career trajectory, the decision to jump scales does not constitute real empowerment. One young interviewee realized that in order to gain access to similar labor market prospects as people on the North Side, she would have to "*become* one of them" and embrace wider-scale identity. She knows that her Lanier area identity is incompatible with a career in the primary labor market. To escape marginality, Lanier area youths must not only jump scales but also deny their place in their own local community.

Conclusion

Conceptualizing ideology as scale contingent enables me to offer a particularly spatial explanation of labor market marginalization of inner-city youths. The wider spatial context of San Antonio frames a dominant ideology that subordinates the neighborhood-based ideology that prevails in the Lanier area. Although young people in the Lanier area strategically select scales of ideological reference for their self-identities, this choice does not constitute an empowering form of agency that provides cultural equality in the labor market. Rather, jumping scale merely facilitates processes of acculturation to dominant cultural norms and rules. Once assimilated, labor market prospects rise for young people. The likely result of not jumping to the wider scale is a career in the secondary labor market. In addition, some young people do not even have the option of jumping scales. Some women, for instance, are locked into a patriarchal family relationship that denies them exposure to external cultural environments and confines them to neighborhood-based ideology.

Although many youths in the Lanier area are painfully aware that neighborhood-based ideologies differ from those that dominate the urban context of San Antonio, such an awareness does not exist among outsiders, who tend to be unfamiliar with the ideology of the Lanier area. Instead, outsiders perceive cultural norms and rules as aspatial and universal, enabling them to represent the Lanier area as dysfunctional space and thereby rationalizing the neighborhood's colonization.

Academic theorization of inner-city marginality often makes a similar mistake. It conceives of ideology as aspatial and accepts universal standards of cultural normality and dysfunction. This practice inscribes identities of social pathology on neighborhoods like the Lanier area and thereby disempowers such communities. If social theorists participate in this process of labeling and marginalizing, then they too legitimize discourses of exclusion, encourage the enforcement of spatial marginalization, and support practices of cultural assimilation. A spatial perspective of ideology is necessary to prevent these consequences.

But what then are possible alternatives? I have argued that jumping scales of ideological reference is more of a process of acculturation than an empowering form of agency that challenges existing power structures. What is needed is a fundamental challenge to the idea that one ideology can legitimately dominate over another. The worldviews of young people in the Lanier area may be different from my own, but they are not dysfunctional. In the conclusion I discuss potential strategies to confront spatial and cultural processes that have pushed inner-city youths to the margin of society and the labor market.

Notes

1. Marston (2000, 219) laments that "questions now driving the scholarship on scale tend to focus on capitalist production while, at best, only tacitly acknowledging and, at worst, outrightly ignoring social reproduction and consumption." Here I examine questions of scale on the social side of the labor market, balancing the existing scholarship on capitalist production.

2. While scale is usually invisible from an internal vantage point, distinct scale-based meanings of places emerge from an outsider perspective. Ley (1977, 508) notes that the "personality of place ranges from the nation state to the neighbourhood church; and habitually interacting groups of people convey a character to the place they occupy which is immediately apparent to an outsider, though unquestioned and taken for granted by habitués."

Conclusion

A risk I took in choosing a supply-side approach to the inner-city labor market lies in oversocializing labor market processes and neglecting the demand side. I realize that supply and demand sides coexist and depend on each other. Not only is the segmentation of labor produced through the cultural judgment of workers, but social inequalities are also constructed through labor market position (Barrera 1979; Fevre 1992). An additional limitation is that the supply side of the labor market is too complex for the scope of a single book. The connection between housing market segregation and labor market segmentation, for example, is crucial for acquiring a comprehensive understanding of the work situation of young minorities in American inner cities.[1] It would be naïve to suggest that any study could provide a complete explanation of the labor market situation of young inner-city Latinos and Latinas. In the end, our social world is more complex and disordered than our compartmentalized approaches to demand and supply sides or housing and labor markets suggest.

Recognizing the epistemological limits to explanation, I cannot offer any complete proposals for rectifying existing problems for young Latinos and Latinas on San Antonio's West Side. Indeed, any suggestion for action may produce unintended consequences, as my critique of the idea of neighborhood effects and its application in academic research and policy making illustrated. But what then should our strategies be for dealing with the problem of labor market exclusion facing inner-city youths? I suggest that we refocus public debate and political activism on normative issues (Galster 1996). Gibson-Graham (1996) has already widened our conception of the "economy" and "class" by treating both notions as discursive constructs. A similar perspective of inner-city labor and urban youths would be a useful response to the unequal treatment of young Lanier area residents in the labor market.

Current discourse of the inner city embraces culture-of-poverty and underclass ideas, which assume that cultural and behavioral deficiencies among inner-city minority youths breed labor attributes that are undesirable in the labor market. In addition, this discourse deploys neighborhood scale to represent dysfunction as a spatially contained problem. This perspective legitimizes, rather than solves, the economic marginalization of neighborhoods and their residents. Furthermore, it conceals labor market processes of cultural exclusion that operate through scale-dependent internal and external representations. The evidence I presented undermines culture-of-poverty and underclass explanations of inner-city marginality. It is possible to intervene in the discourse of the inner-city labor market and to validate internal work identities and affirm place-particular representations of culture. But how should this project of rescripting inner-city labor be accomplished? Who should speak? And what are the concrete policy measures that should drive this process?

A political project aimed at rescripting labor markets must address different spatial scales simultaneously (Harvey 2000). On global, national, and urban scales, labor reward structures have in recent years been realigned in such a manner as to produce increasing polarization between spatial contexts and segments of the labor market. This realignment—related to the global movement of capital and regulation of labor markets—has particular effects on American inner cities: the cultural capital of "secondary" workers is increasingly devalued, while the exchange value of "primary" workers increases (Sassen 1994). The forces of "globalization" and corporate domination have in recent years moved to the top of the agenda among activist groups on the Left, most evident in large-scale demonstrations against regulatory institutions and agreements such as the World Trade Organization, the G8, and Free Trade Agreement of the Americas. Yet, even if these groups win major concessions, the discrepancy between grossly uneven labor reward structures is unlikely to be leveled in the foreseeable future.

On a smaller scale, a system of symbolisms and meanings represents inner-city neighborhoods as marginal space and devalues the cultural capital of young minority and inner-city residents. To maintain this system of labor devaluation amounts to the colonization of inner-city neighborhoods. This system is legitimized by human capital and neighborhood effect theories, promoting the misleading ideas that labor reward structures are entirely market driven, and that inner-city youths are not competitive in the labor market because their neighborhood environment infects them with cultural pathology. But I suggest that discursive cultural processes, rather than the mechanical laws of the market, are responsible for the devaluation of inner-city labor. The discursive nature of neighborhood representations opens the possibility for intervention and for reshaping the discourse of inner-city labor.

Since the value of labor is tied to the cultural judgment of the neighborhood, the revaluation of inner-city labor requires the rescription of the inner-city discourse. A key task in this project is to reconfigure the object of inner-city youths, which currently epitomizes cultural pathology. But to demolish the existing essentialism of inner-city neighborhoods and youths requires more than simply an appreciation of other worldviews. I do not believe that we can overcome our tendency to stereotype and essentialize our surroundings. In fact, we rely on stereotypes and essentialisms to maneuver a world that is too complex for us to grasp in its entirety. We can, however, resort to strategic essentialism and produce new imaginaries that represent inner-city neighborhoods, ethnic minorities, and young people differently. Perhaps we can even move toward "cultural hybridity that entertains difference without an assumed or imposed hierarchy" (Bhabha 1994).

Just as scale has been used to create marginalizing identities, so it can be used to reconstruct these identities. I envision a strategy to empower inner-city communities that promotes the infiltration of marginal identities into dominating ideology. In recent decades, common-law domestic partnerships and queer lifestyles, for example, have increasingly become accepted in mainstream society (although much remains to be accomplished). In this case, outside thinking is shifting toward accepting previously marginalized communities into the realm of the normal.

Similarly, cultural identities of marginal inner-city communities can be "transposed" onto a wider spatial scale. For example, rap music initially became popular in African-American urban neighborhoods and was demonized by the media, mainstream politicians, and middle-class public as deviant and pathological. Yet, over time, rap music has joined other musical styles in being accepted as a dominating musical trend in North America and Europe, legitimizing the musical performances of inner-city youths. In a similar manner, ideas of family, life priorities, and work identity are not necessarily confined to inner-city space but harbor the potential to be incorporated into ideological perspectives of society beyond the neighborhood scale. Spatially contingent meanings of work and labor should be recognized as on a wider scale, and the work identities of inner-city youths should be valorized within the dominating discourse of labor.

The idea of "transposing" identities from a local to a wider scale differs from the notion of "jumping" scale. In Chapter 5 I suggested that individuals jump scales of representation and internalize mainstream norms and values as a strategy of acculturation rather than empowerment. The term "transposing," on the other hand, expresses the process of treating multiple ideological viewpoints as equals and of legitimizing the cultural identity of inner-city youths within dominating society. Young people would no longer need to deny their cultural identity because their neighborhood assumes a subordinate position in the hierarchy of cultural space. Rather, the neighborhood-based community would be treated as a distinct cultural community.

This strategy of opening up the mainstream to alternative articulations of legitimate identities would be assisted by the affirmation of local identity. On a local scale, the inner-city neighborhood functions as a bonding mechanism, and "this bringing together" (Williams 1989, 244) strengthens internal identity and fosters solidarity in the struggle for cultural equality. This process toward community empowerment is already under way in many urban communities in America, and it is driven by local activists and community-based institutions.

The content of current inner-city and labor market discourse, however, constrains the empowerment of local communities. This discourse endows labor market regulators, such as employers (Holzer 1996), with gatekeeping authority over cultural capital, while it removes this authority from inner-city and minority communities. Thus, inner-city discourse constrains any serious challenge to established processes of cultural reproduction that is launched from marginal inner-city space. This situation leaves inner-city residents in a weak position to rescript the inner city.

Community-based institutions are potentially more equipped to derail contemporary popular discourse than marginalized residents. However, these institutions are often responsible for helping young people to overcome immediate labor market barriers, and this responsibility contradicts the long-term project of rescripting the inner city and the labor market. For example, young people in the Lanier area maintain work identities, which are not endorsed outside of the Lanier area. Should community institutions now attempt to reshape young people's labor market identities and influence youths to assimilate to mainstream standards of work? In this case, institutions could be accused of practicing cultural imperialism by denying young people their cultural self-identity and by requesting that they acculturate. Or should institutions support young people's self-defined labor market identities? But in that case, institutional efforts would channel young people into secondary occupations that offer only low wages, few benefits, and limited stability. This dilemma cannot easily be resolved. It relates to established reward structures in the labor market and the common perception that the labor market is a "natural" process. But if the value of labor is set by cultural processes rather than impartial market forces, then it is within the realm of possibilities to emancipate marginal communities, to rewrite the script of the labor market, and to accommodate alternative conceptions of labor. Such a revised script would, for instance, increase the value of the work performed by dishwashers and nursing assistants.

The authority to define the content of the new script should not concentrate among the corporate elite, career politicians, institutional bureaucracies, and academic theorists. Rather, minorities, inner-city communities, and young people must be included in the normative process of valorizing work identities. The issue is not only that diverse groups should have a voice, but also that these voices are heard and enabled to penetrate dominant discourse. Therefore we need channels of communication for stimulating exchange and debate of diverse perspectives, including those currently confined to the realm of cultural illegitimacy. The flow of information in these channels of communication must not be one directional, from the top to the bottom, but circular, to generate fresh imaginaries of work, community, and place. This strategy of reconstructing popular discourse is, of course, not appealing to policy makers, who operate within the boundaries of contemporary discourse.

Nevertheless, public policy does serve an important function of providing ad hoc remedies to unacceptable circumstances of inner-city marginality. Housing policy can be such a remedy. I have argued that low-income housing dispersal and mixed-income housing may merely facilitate acculturation and improve career prospects by bypassing processes of cultural marginalization—not by establishing cultural equality and inclusion. I am not advocating against federal programs, such as Section 8 or Moving to Opportunities, that provide the option to poor families to escape social and economic marginality by moving to the suburbs. If, however, they opt for stigmatized inner-city neighborhoods—which many families indeed prefer—their chances for social upward mobility are severely constrained. Urban enrichment initiatives therefore offer additional ad hoc relief to inner-city communities who opt to stay in urban neighborhoods.

As a midterm strategy, new approaches to residential segregation are necessary. That privileged groups residentially isolate themselves in, say, gated communities illustrates that segregation is not necessarily associated with marginality. For neighborhood-based communities, a degree of residential separation may be desirable to strengthen collective identity and to build social and economic support networks (Dunn 1998; Peach 1996). However, residential segregation must not isolate communities and cut channels of communication. Rather, open communication channels must exist to facilitate exchange between neighborhoods and between different ideological frameworks. Community centers, youth initiatives, and housing programs can play a crucial role and nurture spatial linkages to other communities. Palm Heights institutions are successful in establishing such linkages, but they also tend to privilege external perspectives of work and career over internal perspectives. The aim of building spatial and intercommunity networks should not be the enforcement of acculturation but the demolition of neighborhood stereotypes, false conceptions of dysfunction, and misleading images of dirtiness, aggression, and unreliability. Although this aim seems distant and faint, we cannot afford to lose sight of it.

Notes

1. For examples of studies that address the housing market, see Anderson (1991) and Smith (1989).

Appendix

Technique and Detailed Results of Quantitative Analysis

Principal Component Analysis

As is often the case when working with census data, analysis is limited by the availability of variables pertinent to the problem under study. In the instance of youth labor market marginalization the number of variables available on the tract level based on a consistent age category (eighteen to twenty-four years) was limited to five. A description of these variables is given in table A.1. They represent low educational levels (EDUC12), high poverty levels (POV18-24), and the cross-tabulations of school enrollment and employment (NENR/NEMPL, NENR/EMPL, EN/NEMPL). Theoretically, any combination of nonenrollment and/or nonemployment may indicate marginality in the labor market. The combination of enrolled and employed youths is not included in the analysis, assuming that these youths are not marginalized. Unfortunately, none of the variables distinguishes between men and women.

Table A.1: Variables Included in the Principal Component Analysis

Variable	Description	Mean	S.D.
EDUC12	% of persons who have not completed 12th grade of high school, age 18-24, 1990	30.00	16.59
POV18-24	% of persons who live in poverty, age 18-24, 1990	21.01	15.96
NENR/NEMPL	% of persons who are not enrolled in school and are not employed, ages 18-24, 1990	19.72	12.21
NENR/EMPL	% of persons who are not enrolled in school and are employed, ages 18-24, 1990	35.09	12.65
EN/NEMPL	% of persons who are enrolled in school and are not employed, ages 18-24, 1990	19.01	10.30

Source: U.S. Bureau of the Census, 1995

A principal component analysis extracted two underlying dimensions from the five variables. The component matrix was then rotated using a Varimax rotation. The results are presented in table A.2. The two factors account for 74.44 percent of the total variance among the five variables. The first factor explains 47.68 percent of the variance; the second factor explains 26.76 percent. An Oblim rotation tested for true orthogonality of the two factors by means of relaxing the condition of orthogonality. The small interfactor correlation of -.128 indicates true orthogonality between the two factors.

Table A.2: Results of Principal Component Analysis

Label	Factor 1 Marginal Youth	Factor 2 Employed/Enrolled Youth
Percent of Explained Variance	47.68	26.76
Component Matrix (Varimax Rotation)		
EDUC12	.889*	.160
POV18-24	.812*	.174
NENR/NEMPL	.922*	.093
NENR/EMPL	.131	-.826*
EN/NEMPL	-.358	.728*

* Indicates high commonalties (factor scores < -.6 or > .6)

The factor loadings, indicating the correlation of each variable with the two underlying factors, are used to interpret and label the factors. The first factor is labeled Marginalized Youths; the second, Employed/Enrolled Youth. The latter differentiates tracts according to a bipolar contrast between the variables measuring youths who are neither enrolled in school nor employed, and youths who are both enrolled in school and employed. The mapping analysis projected the scores of the factor Marginalized Youths onto a census tract map of Bexar County.

Regression Analysis

A stepwise least square regression analysis was performed on the factor Marginalized Youths as the dependent variable and ten independent variables (table A.3). The first set of variables measures selected behavioral characteristics within a census tract: mothers below seventeen years of age (BIRTH17); births by single mothers (BIRTHSINGLE); and the percentage of women who were never married (MARRIED). Behavior that is represented by these variables does not indicate pathology, but rather the effects of limited educational and employment opportunities available to single and teenage mothers. Although the variables represent women only, Wilson (1987) suggests that they are related to behavior and employment attributes of local men who, if unemployed, are not marriage candidates. Whether one accepts Wilson's explanation or not, these variables do represent distinct perspectives on marriage and family prevalent in census tracts. These measures of youth behavior are expected to be positively correlated with the marginalization of youth.

Table A.3: Variables by Census Tract Considered for Regression Analysis

Variable	Description	Mean	S.D.
Behavioral Characteristics			
BIRTH17[1,2]	% of births by mothers under 17 years, all ages	3.48	2.97
BIRTHSINGLE[1]	% of births by single mothers, all ages	17.59	11.64
MARRIED	% of women never married, ages 15 to 24	24.50	11.14
Labor Market Attributes			
HHINC	Mean household income	*28205*	*20544*
POVERTY[2]	% of persons in poverty, all ages	12.99	11.95
INDPRIM	% of employed persons in independent primary labor market segment, all ages	26.31	14.49
SUBPRIM	% of employed persons in subordinate primary labor market segment, all ages	49.60	7.50
SECOND	% of employed persons in secondary labor market segment, all ages	24.09	13.09
NOWORK	% of persons who did not work, ages 16 and older	31.85	12.39
PART	% of persons who worked less than 35 hrs/week, 16 and older	10.53	2.97
TEMP	% of persons who worked less than 40 weeks, ages 16 and older	14.64	4.01
Ethnic and Origin Characteristics			
FOREIGN	% of persons foreign born, all ages	8.94	7.86
HISPANIC	% of persons who are Hispanic	47.39	29.68
LINGISOL[2]	% of persons linguistically isolated, all ages	8.61	8.97

Sources: U.S. Bureau of the Census (1995); San Antonio Metropolitan Health District (1994)
[1] Data for 1993
[2] Variable was dropped in the analysis due to skewedness and autocollinearity

A second set of neighborhood variables measure employment attributes of the adult population. These include household income (HHINC) and overall poverty levels (POVERTY). Occupational composition of a tract's workforce is measured by dividing the labor market into three occupational segments using the scheme in table A.4. The three segments are represented by the variables INDPRIM, SUBPRIM, SECOND. Additional labor market information is provided by the percentage of persons who are not working (NOWORK), who are working part-time (PART), and who engage in temporary work (TEMP). Adults who are not employed on a full-time basis may lower career prospects for youths. If the characteristics of adults in the neighborhood affect youths, then overall employment characteristics of a tract's adult workers should correlate with the factor Marginalized Youths.

Table A.4: Categorization of Occupations into Labor Market Segments

Labor Market Segment	U.S. Census Occupational Category (Code)
Independent Primary	Executive, Administrative, and Managerial Occupations (0-42)
	Professional Specialty Occupations (43-202)
	Protective Service Occupations (413-432)
Subordinate Primary	Technicians and Related Support Occupations (203-242)
	Sales Occupations (243-302)
	Administrative Support Occupations, including Clerical Service Occupations (303-402)
	Farming, Forestry, and Fishing Occupations (473-502)
	Precision Production, Craft, and Repair Occupations (473-702)
	Transportation and Material Moving Occupations (803-863)
Secondary	Private Household Occupations (403-412);
	Service Occupations, except protective and household (433-472)
	Machine Operators, Assemblers, and Inspectors (703-802)
	Handlers, Equipment Cleaners, Helpers, and Laborers (864-999)

Based on Gittleman and Howell, 1995

The final three variables in the regression model measure ethnic and origin characteristics of census tracts: the percentage of persons foreign-born (FOREIGN); the percentage of Hispanics (HISPANIC); and the percentage of persons who live in households in which at least one person does not speak English, i.e., linguistic isolation (LINGISLO). If origin and ethnic discrimination create a labor market barrier for young people, then these three variables should predict youth labor market marginalization.

Three of the variables originally considered for the analysis (see table A.3) were not included in the model because of their nonnormal distributions (BIRTH17) or multicollinearity (POVERTY, LINGISOL) with other variables. The variable INDPRIM was included in the regression model but was removed by the stepwise regression procedure. The regression results are displayed in table A.5. An analysis of residuals indicated a normal distribution. Furthermore, neither a plot of residuals against the factor Marginalized Youths nor the mapping of residuals by census tract revealed any remaining systematic regularities in the data.

Table A.5: Ordinary Least Square Regression Results for Marginal Youth

Variable	ß-Coefficient	t
Behavioral Characteristics		
BIRTHSINGLE	2.925**	6.306
MARRIED	.722*	2.342
Labor Market Attributes		
HHINC	.022	0.633
SUBPRIM	-.360	-0.606
SECOND	2.220**	3.791
NOWORK	1.546**	3.045
PART	-3.399**	-2.626
TEMP	3.993**	4.079
Ethnic/Cultural Characteristics		
FOREIGN	1.558	1.484
HISPANIC	.422*	1.991

Multiple R = .914
R^2 = .835
Model F = 102.000**
* Significant at .05
** Significant at .01

Case Study Area Selection

The selection of the two case study areas is based on census tract factor scores for the Marginalized Youths factor. Most tracts with high factor scores also have a relatively large Hispanic population. The two tracts representing the Lanier area were selected as primary study area because of the *high* factor scores of 3.75 (tract 1105) and 1.98 (tract 1702). I chose Palm Heights as the control area on the basis of *low* factor scores of -.04 (tract 1504) and -2.3 (tract 1602). What makes the two areas appealing for a comparative study is their relatively low socioeconomic status, large Hispanic population, and similar location in the context of greater San Antonio.

Method of Qualitative Analysis

I made an effort to conduct at least one interview at each community-based organization that provides career and education related services to youth in the Lanier area and in Palm Heights. The organizations were identified through a variety of sources, including the San Antonio Business Listings, Yellow Pages, referrals by other organizations, and word of mouth. Two institutions decided not to participate in the survey. An issue was that some institutions provide services that are only indirectly related to youth career development and education. If indirect linkages between an institution and youth career development were identified, an institution was included in the survey. In some cases, I was able to observe institutional service provision in action.

Table A.6 lists the institutions, services provided, target population, number of clients, and service areas. Six institutions service youth in the Lanier area, three institutions service Palm Heights, and eight institutions service both areas. The Lanier area has more institutions that offer a greater variety of education and employment related services than Palm Heights. Unfortunately, the exact numbers of youths who live within the boundaries of the two study areas and who are involved with the institutions are unavailable. However, the large number of clients who are serviced by the institutions suggests that a high proportion of youths have contact with institutions. For instance, Inner City Development serves up to 400 youth in the Lanier area, and the two tracts representing this area contained 1049 youth between the ages of eighteen and twenty-four, according to the 1990 census. Catchment areas are inconsistent across institutions, making it difficult to link institutions with the particular youth populations in the two study areas. To deal with this issue interviews were structured in such a way that the discussion focused on youths who live in the two study areas. For instance, interview respondents were asked to comment on their experiences with youths in the Lanier area and/or Palm Heights. The interviews focused on institutional perceptions of the neighborhood, services provided, and institutional practices toward career development for youths. This concern is broader than the success rate of individual training and placement programs and encompasses normative and ideological issues.

Table A.6: Institutional Profiles

Institution	Target Population	Number of Clients	Service Area	Education, Job, Training, & Placement	Other
Lanier Area					
1. S. Lanier Band Booster Assoc.	Grades 9-12	40 students	Lanier High School		Mentoring
2. Family Self Sufficiency	Above age 16	152 families	Public Housing		Referrals
3. Downtown Youth Center	Ages 6-18	30-55 youths	West Side	Tutoring	Recreation
4. Inner City Development	All local youths	Up to 400 youths	Lanier Area	Tutoring; volunteering	Recreation
5. Young Life	Ages 6-18	50-80 youths	West Side		Recreation Counseling
6. Guadalupe Church	Ages 13-24	20-30 youths	West Side	Counseling	Mentoring
Palm Heights					
7. Joven	Ages 10-17, "at risk"	250-300 youths	South Side		Case management
8. Palm Heights Rec. Center	Ages 6-19	250 youths	South Side		Recreation
9. Saint John's Catholic Church	Ages 12-20	40-60 youths	Palm Heights	Volunteering	Mentoring
Lanier Area & Palm Heights					
10. Positive Solutions	Ages 14-21, low income	800-2,000 students	South & West Sides	GED; training; placement	Counseling
11. Literacy Service Division	Above age 17	n.a.	South & West Sides	GED	Literacy
12. S.A. Education Partnership	Grades 9-12	n.a.	South & West Sides	Scholarships	Counseling
13. S.E.R. Jobs for Progress	Ages 14-18	40 youths	South & West Sides	GED; training; placement	
14. Juvenile Probation Office	Students under age 18	200 students/week	South & West Sides		Probation review
15. Project Stay	Grade 12, low income	425 students	San Antonio	College placement	Referrals
16. YWCA	Ages 13-24	500 youths	San Antonio	Counseling; volunteering	Child care; parenting
17. TX Workforce Commission	Welfare recipients	320	Bexar County	GED; training; placement	Case management

Sources: based on interview data (1996)

The sample for youths was obtained through community service providers, local employers, and snowball sampling. A stratified sampling technique ensured that respondents varied in gender, education, skills, and employment characteristics. Table A.7 displays the interview profiles for youths. The interviews dealt with the formation of career identities and how these identities relate to gender, ethnicity, age, and neighborhood circumstances. The interviewing process can be an intimidating experience for young respondents, so I attempted to make interviewees as comfortable as possible. I asked them before the interview how they feel about the interview being audiotaped. If they disagreed, notes were taken. Most respondents, however, agreed with audiotaping. In addition, I assured respondents of strict anonymity and guaranteed that no third party has access to the tapes or the notes.

Separate interview guides were developed for youths and institutions. Interviews followed an interviewing technique called focused interviewing (Zeisel 1981). According to this technique, I could divert from the structured interview guide and ask respondents to elaborate on certain issues that required clarification or that I did not anticipate prior to the interview. As a result of this semistructured technique, interviews sometimes developed into less-formal conversations, which put many respondents at ease. This technique also provided flexibility in order to address epistemological issues. For instance, institutional administrators and young members of peer groups often performed the roles expected from them, especially when they were audiotaped. By modifying the interview guides and adjusting the interviewing technique, I was able to examine the nature of these roles. In addition, I was able to explore and (re)consider my own positionality as the observer and my own ideological preconceptions and resources for interpreting spatial and social relationships (Rose 1997).

The different interview sample sizes in the two study areas relate to a sequential research design. I first examined processes of career identity formation in the Lanier area; once these processes were analyzed, Palm Heights was investigated for cross reference to make a counterpoint. I did not intend to replicate the entire Lanier area study in Palm Heights, but rather examine the processes of career identity formation observed in the Lanier area for their place specificity. Thus, fewer interviews were necessary in Palm Heights.

Table A.7: Profile of Youths Interviewed

	Lanier Area	Palm Heights Area
Total	21	8
Men	9	4
Women	12	4
Hispanic/Latino	21	8
Age		
16–17	11	8
18–21	8	0
22 and above	2	0
Employment		
Working full-time	4	0
Working part-time	6	2
Not working	11	6
Income less $7.00/hour	7	2
Looking for work	7	3
Education		
In school/college	14	8
Not in school/college	7	0
Not working and not in school	2	0
Family		
Interviewee has no children	13	8
Interviewee has one child	5	0
Interviewee two children	3	0
Spouse/partner in household	2	0
Interviewee receives AFDC	3	0
Family Background		
Raised by both parents	9	6
Raised by one parent only	8	2
Raised by others, other than parents	3	0
Parents receive AFDC	7	1
Parent(s)/guardian(s) work	16	7
No parent/guardian works	5	1
Language of Interview		
English	21	8
Spanish	0	0

The relatively small sample sizes suited the in-depth method of analysis. I used a variation of grounded theory analysis whereby data collection, analysis, and theory building temporally overlapped (Strauss 1987). This approach fit the exploratory nature of this research. It enabled me to channel the theoretical insights gained from the empirical data immediately back into the research process and adjust the data collection and interviewing procedures accordingly (Burgess 1984, 143-165; Silverman 1985, 22). Thus I could cross-verify initial conclusions drawn from the data while the data-collection process was still ongoing. In addition, I gained important insights through participant observation in my role as a volunteer for a youth program in the Lanier area and daily interaction with local residents. These insights were crucial for developing an appreciation for local and community-particular perspectives and for issues that to me, as an outsider, were not immediately apparent.

Bibliography

Abramson, Alan J., and Michael E. Fix. 1993. Growth without prosperity reveals surprising trends. *Partnership for Hope Newsletter* (San Antonio) 3 (2):1-5.

Abramson, Alan J., Mitchell S. Tobin, and Matthew R. VanderGoot. 1995. The changing geography of metropolitan opportunity: The segregation of the poor in U.S. metropolitan areas, 1970 to 1990. *Housing Policy Debate* 6 (1): 45-72.

Agnew, John A. 1987. *Place and politics: The geographical mediation of state and society*. Boston: Allen & Unwin.

Aitken, Stuart C. 2000. Mothers, communities and the scale of difference. *Social and Cultural Geography* 1 (1): 65-82.

Anderson, Elijah. 1999. *Code of the street: Decency, violence, and the moral life of the inner city*. New York: Norton.

Anderson, Kay J. 1991. *Vancouver's Chinatown: Racial discourse in Canada, 1875-1980*. Montreal: McGill-Queen's University Press.

Averitt, Richard T. 1968. *The dual economy: The dynamics of American industrial structure*. New York: Morton.

Barrera, Mario. 1979. *Race and class in the Southwest: A theory of racial inequality*. South Bend, IN: University of Notre Dame Press.

Bauder, Harald. 2000. Reflections on the spatial mismatch debate. *Journal of Planning Education and Research* 19 (3): 316-20.

Bhabha, Homi K. 1994. *The location of culture*. London: Routledge.

Bohl, Charles C. 2000. New urbanism and the city: Potential applications and implications for distressed inner-city neighborhoods. *Housing Policy Debate* 11 (4): 761-801.

Bourdieu, Pierre. 1984. *Distinction: A social critique of the judgment of taste*. Cambridge, Mass.: Harvard University Press.

Bourdieu, Pierre, and Jean-Claude Passeron. 1977. *Reproduction in Education, Society and Culture*. London: Sage.

Bourgois, Phillippe I. 1995. *In search of respect: Selling crack in El Barrio*. Cambridge, N.Y.: Cambridge University Press.

Bowditch, Christine. 1993. Getting rid of troublemakers: High school disciplinary procedures and the production of dropouts. *Social Problems* 40 (4): 493-509.

Bowlby, Sophie, Sally Loyed Evans, and Robina Mohammad. 1998. The workplace—Becoming a paid worker: Image and identity. In *Cool places: Geographies of youth cultures*, edited by Tracey Skelton and Gill Valentine, 229-48. London: Routledge.

Briggs, Xavier de Souza, Joe T. Darden, and Angela Aidala. 1999. In the wake of desegregation: Early impacts of scattered-site public housing on neighborhoods in Yonkers, New York. *Journal of the American Planning Association* 65 (1): 27-49.

Brooks-Gunn, Jeanne, Greg J. Duncan, and J. Lawrence Aber, eds. 1997. *Neighborhood poverty: Volume II—Policy implications in studying neighborhoods*. New York: Russell Sage.

Browning, Jeremy D. 1994. The reproduction of racial differences in educational achievement: A structurationist achievement. In *Marginalized places and populations: A structurationist agenda*, edited by David Wilson and James O. Huff, 165-76. Westport, Conn.: Praeger.

Burgess, Robert G. 1984. *In the field: An introduction to field research*. London: George Allen Unwin.

Capps, Randy. 1996. Quality, not quantity of jobs is prime concern for San Antonio work force. *Partnership for Hope Newsletter* (San Antonio) 6 (1):1, 4-7.

Cardenas, Gilberto, Jorge Chapa, and Susan Burek. 1993. The changing economic position of Mexican Americans in San Antonio. In *Latinos in a Changing U.S. Economy*, edited by R. Morales and F. Bonilla, 160-83. Newbury Park, Calif.: Sage.

Chiricos, Ted, Ranee McEntire, and Mard Gertz. 2001. Perceived racial and ethnic composition of neighborhood and perceived risk of crime. *Social Problems* 48 (3): 322-40.

Cisneros, Henry G. 1995. *Regionalism: The new geography of opportunity*. Washington D.C.: U.S. Department of Housing and Urban Development.

Cityscape. 1997. Mixed income housing: In memory of Donald Terner. Special issue, 3 (2).

Cope, Meghan. 1998. Home-work links, labor markets, and the construction of place in Lawrence, Massachusetts, 1920-1939. *Professional Geographer* 50 (1): 126-40.

Coulton, Claudia J., Jill E. Korbin, and Marilyn Su. 1999. Neighborhoods and child maltreatment: A multilevel study. *Child Abuse & Neglect*, 32 (11): 1019-40.

Crane, Jonathan. 1991a. Effects of neighborhoods on dropping out of school and teenage childbearing. In *The urban underclass*, edited by Christopher Jencks and Paul E. Peterson, 299-320. Washington, D.C.: Brookings Institution:

———. 1991b. The epidemic theory of ghettos and neighborhood effects on dropping out and teenage childbearing. *American Journal of Sociology* 96: 1226-59.

Crang, Philip. 1994. It's showtime: On the workplace geographies of display in a restaurant in southeast England. *Environment and Planning* 12: 675-704.

Cross, Harry, Genevieve Kenney, Jane Mell, and Wendy Zimmerman. 1990. *Employer hiring practices: Differential treatment of Hispanic and Anglo job seekers*. Urban Institute Report 90-4. Washington, D.C.: Urban Institute.

Dalaney, D., and H. Leitner. 1997. The political construction of scale. *Political Geography* 16 (2): 93-97.

Davis, Mike. 1990. *City of quartz: Excavating the future of Los Angeles*. London: Verso.

De Oliver, Miguel. 1996. Historical preservation and identity: The Alamo and the production of a consumer landscape. *Antipode* 28 (1): 1-23.

———. 2001. Multicultural consumerism and racial hierarchy: A case study of market culture and the structural harmonization of contradictory doctrines. *Antipode* 33: 229-59.

Dunn, Kevin M. 1998. Rethinking ethnic concentration: The case of Cabramatta, Sidney. *Urban Studies*. 35 (3): 503-27.

Edin, Kathryn, and Laura Lein. 1997. *Making ends meet: How single mothers survive welfare and low wage work*. New York: Russell Sage Foundation.

England, Kim V. L. 1993. Suburban pink collar ghettos: The spatial entrapment of women. *Annals of the Association of American Geographers* 83: 225-42.

Evans, William N., Wallace E. Oates, and Robert M. Schwab. 1993. Measuring peer group effects: A study of teenage behavior. *Journal of Political Economy* 100 (5): 966-91.

Fainstein, Norman. 1993. Race, class, and segregation: Discourses about African-Americans. *International Journal of Urban and Regional Research* 17: 384-403.

Fernández Kelly, Patricia M. 1994. Towanda's triumph: Social and cultural capital in the transition to adulthood in the urban ghetto. *International Journal of Urban and Regional Research* 18: 88-111.

Fernandez, Roberto, and David Harris. 1992. Social isolation and the underclass. In *Drugs, crime and social isolation: Barriers to urban opportunity*, edited by Adele V. Harrell and George E. Peterson, 257-93. Washington, D.C.: Urban Institute.

Fevre, Ralph. 1992. *The sociology of labor markets*. New York: Harvester Wheatsheaf.

Foucault, Michel. 1970. *The order of things: An archaeology of the human sciences*. New York: Pantheon Books.

Furstenberg, Frank F., Thomas D. Cook, Jacquelynne Eccles, Glen H. Elder, Jr., and Arnold Sameroff. 1999. *Managing to Make It: Urban Families and Adolescent Success*. Chicago: University of Chicago Press.

Galster, George C. 1996. Urban issues, policies, and research: Examining the interface. In *Reality and Research: Social Science and U.S. Urban Policy since 1960*, edited by G. Galster. Washington, D.C.: Urban Institute.

Galster, George C., and Maris Mikelsons. 1995. The geography of metropolitan opportunity: A case study of neighborhood conditions confronting youth in Washington, D.C.. *Housing Policy Debate* 6 (1): 73-102.

Galster, George C., and Sean P. Killen. 1995. The geography of metropolitan opportunity: A reconnaissance and conceptual framework. *Housing Policy Debate* 6 (1): 7-43.

Gans, Herbert J. 1990. Deconstructing the underclass: The term's dangers as a planning concept. *Journal of the American Planning Association*. 56 (3): 271-77.

Gephart, M. A. 1997. Neighborhoods and communities as contexts for development. In *Neighborhood poverty—Volume 1: Context and consequences for children*, edited by Jeanne Brooks-Gunn, Greg J. Duncan and J. Lawrence Aber, 1-43. New York: Russell Sage Foundation.

Gibson-Graham, J. K. 1996. *The end of capitalism (as we knew it): A feminist critique of political economy*. Malden, Mass.: Blackwell.

Giddens, Anthony. 1984. *The constitution of society: Outline of the theory of structuration*. Berkeley: University of California Press.

Gilbert, Melissa R. 1998. "Race," space, and power: The survival strategies of working poor women. *Annals of the Association of American Geographers* 88 (4): 595-621.

Gittleman, Maury B., and David R. Howell. 1995. Changes in the structure and quality of jobs in the United States: Effects by race and gender. *Industrial and Labor Relations Review* 48 (3): 420-40.

Goldberg, Carey. 1997. Hispanic households struggle as poorest of U.S. poor. *New York Times*, 30 January.

Gordon, David M., Richard Edwards, and Michael Reich. 1982. *Segmented work, divided workers: The historical transformation of labour in the United States*. Cambridge, N.Y.: Cambridge University Press.

Gregory, Derek. 1978. *Ideology, science and human geography*. London: Hutchinson.

———. 1994. *Geographical imaginations*. Cambridge, Mass.: Blackwell.

———. 2000. Epistemology. In *The dictionary of human geography*, 4th edition, edited by R. J. Johnston, D. Gregory, G. Pratt, and M. Watts, 226-28. Oxford, Eng.: Blackwell

Hanson, Susan, and Geraldine Pratt. 1995. *Gender, work, and space*. London: Routledge.

Harris, Fred R., and Lynn A. Curtis. 1998. *Locked in the poorhouse: Cities, race and poverty in the United States*. Lanham, Md.: Rowman & Littlefield.

Harrison, Bennett, and Marcus S. Weiss. 1998. *Workforce development networks: Community-based organizations and regional alliances*. Thousand Oaks, Calif.: Sage.

Harvey, David. 1990. *The condition of postmodernity: An enquiry into the origins of cultural change*. Cambridge, Mass.: Blackwell.
———. 2000. *Spaces of hope*. Berkeley: University of California Press.
Herod, Andrew. 1991. The production of scale in United States labor relations. *Area* 23 (1): 82-88.
Holloway, Sarah L. 1999. Mother and Worker? The negotiation of motherhood and paid employment in two urban neighborhoods, *Urban Geography* 20 (5): 438-60.
Holzer, Harry J. 1991. The mismatch hypothesis: What has the evidence shown? *Urban Studies* 28: 105-22.
———. 1996. *What employers want: Job prospects for less-educated workers*. New York: Russell Sage Foundation.
Hopgood, Mei-Ling. 2000. Complaint cites bias by a temp firm agency: It illegally denies jobs to some, employee says. *Detroit Free Press*, 5 April, 1A, 11A.
Hughes, Mark Alan. 1989. Misspeaking truth to power: A geographical perspective on the "underclass" fallacy. *Economic Geography* 65 (3): 187-207.
Ihlanfeldt, Keith R., and David L. Sjoquist. 1998. The spatial mismatch hypothesis: A review of recent studies and their implications for welfare reform. *Housing Policy Debate* 9 (4): 849-92.
Irwin, Sarah. 1995. Social reproduction and change in the transition from youth to adulthood. *Sociology* 29 (2): 293-315.
Jackson, Peter. 1991. Mapping meanings: A cultural critique of locality studies. *Environment and Planning A* 23: 215-28.
Jankowski, Martín Sánches. 1986. *City bound: Urban life and political attitudes among Chicano youth*. Albuquerque: University of New Mexico Press.
Jargowsky, Paul A. 1997. *Poverty and place: Ghettos, barrios, and the American city*. New York: Russell Sage Foundation.
Jencks, Christopher, and Susan E. Mayer. 1990. The social consequences of growing up in a poor neighborhood. In *Inner city poverty in the United States*, edited by L. E. Lynn, Jr., and M. G. H. McGeary, 111-86. Washington D.C.: National Academy Press.
Johnson David R., John A. Booth, and Richard J Harris. 1983. *The politics of San Antonio: Community, progress and power*. Lincoln: University of Nebraska Press.
Kain, John F. 1992. The spatial mismatch hypothesis: Three decades later. *Housing Policy Debate* 3 (2): 371-460.
Kasarda, John D. 1990. The jobs-skills mismatch. *New Perspectives Quarterly* 7 (4): 34-7.
———. 1993. Inner-city poverty and economic access. In *Rediscovering urban America: Perspectives on the 1980s*, edited by Jack Sommer and Donald A. Hicks, 4-1 to 4-60. Washington, D.C.: U.S. Department of Housing and Urban Development.

Kasinitz, Philip, and Jan Rosenberg. 1996. Missing the connection: Social isolation and employment on the Brooklyn Waterfront. *Social Problems* 43 (2): 180-97.
Kelly, Robin D. G. 1997. *Yo' Mama's disfunktional! Fighting the Culture Wars in Urban America*. Boston, Mass.: Beacon.
Kenrick, Jane. 1981. Politics and the construction of women as second-class workers. In *The Dynamics of Labour Market Segmentation*, edited by Frank Wilkinson, 167-91. London: Academic Press.
Kessler-Harris, Alice. 1982. *Out to work: A history of wage-earning women in the United States*. New York: Oxford University Press.
Kett, Joseph F. 1977. *Rites of passage: Adolescence in America— 1790 to the present*. New York: Basic.
Kirschenman, Joleen, and Katheryn M. Neckerman. 1991. 'We'd love to hire them, but. . . .' The Meaning of Race for Employers. In *The urban underclass*, edited by Christopher Jencks and Paul E. Peterson, 203-32. Washington, D.C.: Brookings Institution.
Konstam, Patricia. 1996a. S.A. sees lowest jobless rate ever. *San Antonio Express-News*, 20 Nov., A1, A8.
———. 1996b. Job skills targeted in $863,887 grant. *San Antonio Express-News*, 25 Oct., B1, B3.
Kozol, Jonathan. 1991. *Savage inequalities: Children in America's schools*. New York: Harper.
Kwan, Mei-Po. 1999. Gender and individual access to urban opportunities: A study using space-time measures. *Professional Geographer* 51 (2): 210-27.
Lee, Gloria, and John Wrench. 1987. Race and gender dimensions of the youth labour market: From apprenticeship to YTS. In *The manufacture of disadvantage: Stigma and social closure*, edited by Gloria Lee and Ray Loveridge, 83-99. Milton Keyes, Eng.: Open University Press.
Lefebvre, Henri. 1991. *The production of space*. Translated by Donald Nicholson-Smith. Oxford, Eng.: Blackwell.
Lewis, Oscar. 1965. *La Vida: A Puerto Rican Family in the Culture of Poverty—San Juan and New York*. New York: Vintage Books/Random House.
———. 1966. The culture of poverty. *Scientific American* 215 (4): 19-25.
Ley, David. 1974. *The black inner city as frontier outpost: Images and behavior of a Philadelphia neighborhood*. Washington D.C.: The Association of American Geographers.
———. 1977. Social geography and the taken-for-granted world. *Transactions of the Institute of British Geographers* 2: 498-51.
———. 1996. *The new middle class and the remaking of the central city*. New York: Oxford University Press.
Longhurst, B. 1991. Raymond Williams and local cultures. *Environment and Planning A* 13: 229-38.
Marsden, David. 1986. *The end of economic man? Custom and competition in labour markets*. New York: St. Martin's.

Marston, Sallie A. 2000. The social construction of scale. *Progress in Human Geography* 24 (2): 219-42.
Massey, Doreen. 1984. *Spatial division of labour: Social structures and the geography of production.* New York: Macmillan.
Mattingly, Doreen J. 1999. Job search, social networks, and local labor market dynamics: The case of paid household work in San Diego, California. *Urban Geography* 20 (1): 46-74.
McDowell, Linda. 1997. *Capital culture: Gender at work in the city.* Oxford, Eng.: Blackwell.
Menchaca, Martha. 1995. *Mexican outsiders: A community history of marginalization and discrimination in California.* Austin: University of Texas Press.
Miller, Byron A. 2000. *Geography and social movements: Comparing antinuclear activism in the Boston area.* Minneapolis: University of Minnesota Press.
Mirriam-Webster's Dictionary, 10th edition. 2002. Springfield, Mass.: Mirriam-Webster, Inc. <www.m-w.com>.
Mitchell, Don. 1995. There's no such thing as culture: Toward a reconceptualization of the idea of culture in geography. *Transactions of the Institute of British Geographers* 20, 102-16.
Montejano, David. 1987. *Anglos and Mexicans in the making of Texas, 1836-1986.* Austin: University of Texas Press.
Morales, Rebecca, and Frank Bonilla. 1993. Restructuring and the new inequality. In *Latinos in a changing U.S. economy: Comparative perspectives on growing inequality*, edited by Rebecca Morales and Frank Bonilla, 1-27. Newbury Park, Calif.: Sage.
Moynihan, Daniel P. 1965. *The Negro family: The case for national action.* Washington, D.C.: Office of Policy Planning and Research, U.S. Department of Labor.
Murray, C. 1984. *Losing ground: American social policy 1950-1980.* New York: Basic.
Nenno, Mary K. 1998. New directions for federally assisted housing: An agenda for the Department of Housing and Urban Development. In *New Directions in Urban Public Policy*, edited by David P. Varady, Wolfgang F. E. Preiser, and Francis P. Russell, 205-25. New Brunswick, N.J.: Center for Urban Policy Research.
Newman, Katherine S. 1999. *No shame in my game: The working poor in the inner city.* New York: Alfred A. Knopf and the Russell Sage Foundation.
Offe, Claus, and Karl Hinrichs. 1985. The political economy of the labour market. In *Disorganized capitalism: Contemporary transformations of work and* politics, edited by Claus Offe, 10-51. Cambridge, Mass.: Polity.
Omi, Michael, and Howard Winant. 1986. *Racial formation in the United States: From the 1960s to the 1980s.* New York: Routledge.

O'Regan, Katherine M., and John M. Quigley. 1996. Spatial effects upon employment outcomes: The case of New Jersey teenagers. *New England Economic Review: Special issue, Earnings equality*: 41-64.

Parkin, Frank. 1974. Strategies of social closure in class formation. In *The Social Analysis of Class Structure*, edited by Frank Parkin. London: Tavistock.

Partnership for Hope. 1993. Poor in San Antonio have few housing options. *Partnership for Hope Newsletter* (San Antonio) 3 (3).

Pattillo-McCoy, Mary. 1999. *Black picket fences: Privilege and peril among the Black middle class*. Chicago: University of Chicago Press.

Peach, Ceri. 1996. Good segregation, bad segregation. *Transactions of the Institute of British Geographers* 21: 216-35.

Peck, Jamie A. 1996. *Work-place: The social regulation of labor markets*. New York: Guilford.

Pendall, Rolf. 2000. Why Voucher and Certificate Users Live in Distressed Neighborhoods. *Housing Policy Debate* 11 (4): 881-910.

Pile, Steve, and Nigel Thrift. 1995. Introduction. In *Mapping the subject: Geographies of cultural transformation*, edited by Steve Pile and Nigel Thrift, 1-12. London: Routledge.

Pinderhughes, Howard. 1997. *Race in the hood: Conflicts and violence among urban youth*. Minneapolis: University of Minnesota Press.

Pratt, Geraldine. 1989. Reproduction, class, and the spatial structure of the city. In *New models in geography: The political economy perspective*. Vol. 2, edited by Richard Peet and Nigel Thrift. London: Unwin Hyman.

Preston, Valerie, and Sarah McLafferty. 1999. Spatial mismatch research in the 1990s: Progress and potential. *Papers in Regional Science* 78: 387-402.

Reich, Michael, David M. Gordon, and Richard C. Edwards. 1973. Dual labor markets: A theory of labor market segmentation. *American Economic Review* 63 (2): 359-65.

Ricketts, Erol R., and Isabel V. Sawhill. 1988. Defining and measuring the underclass. *Journal of Policy Analysis and Management* 7 (2): 316-25.

Rosales, Rodolfo. 1999. Personality and Style in San Antonio Politics. In *Chicano politics and society in the late Twentieth Century*, edited by D. Montejano. Austin: University of Texas Press.

Rose, Gillian. 1997. Situating knowledges: Positionality, reflexivities and other tactics. *Progress in Human Geography* 21: 305-20.

Rosenbaum, Emily, and Laura E. Harris. 2001. Residential mobility and opportunities: Early impacts on the moving to opportunity demonstration program in Chicago. *Housing Policy Debate* 12 (2): 321-46.

Rosenbaum, James F. 1991. Black pioneers—do their moves to the suburbs increase economic opportunity for mothers and children? *Housing Policy Debate* 2 (4): 1179-213.

———. 1995. Changing the geography of opportunity by expanding residential choice: Lessons from the Gautreaux Program. *Housing Policy Debate* 6 (1): 231-69.

Ruddick, Susan M. 1996. *Young and homeless in Hollywood: Mapping social identities*. New York: Routledge.
Sampson, Robert J., Stephen W. Raudenbush, and Felton Earls. 1997. Neighborhoods and violent crime: A multilevel study of collective efficacy. *Science* 227: 918-24.
San Antonio Metropolitan Health District. 1994. *1993 Bexar County census tract data sets: Maternal health indicators, teen births and births by hospital; deaths by selected causes*. San Antonio, Tex.
Sassen, Saskia. 1994. *Cities in a world economy*. Thousand Oaks, Calif.: Pine Forge.
Sayer, Andrew. 1984. *Method in social science: A realist approach*. London: Hutchinson.
Schwab, William A. 1992. *The sociology of cities*. Englewood Cliffs, N.J: Prentice Hall.
Segura, Denis. 1995. Labor market stratification: The Chicana experience. In *Latina issues: Fragments of historia(ella) (herstory)*, edited by Antoinette Sedillo López. New York: Garland Publishing.
Shuttles, Gerald D. 1968. *The social order of the slum: ethnicity and territory in the inner city*. Chicago: The University of Chicago Press.
Sibley, David. 1995. *Geographies of exclusion: Society and difference in the West*. London: Routledge.
———, ed. 1998. Theme issue: "Social exclusion." *Geoforum* 29 (2): 119-206.
Silverman, David. 1985. *Qualitative methodology and sociology: Describing the social world*. Aldershot, Eng.: Gower Publishing.
Simcha-Fagan, Ora, and Schwarz Joseph E. 1986. Neighborhood and delinquency: An assessment of contextual effects, *Criminology*, 24 (4): 667-704.
Skinner, Curtis. 1995. Urban labor markets and young black men: A literature review. *Journal of Economic Issues* 29 (1): 47-65.
Smith, Neil. 1984. *Uneven development: Nature, capital and the production of space*. Oxford, Eng.: Blackwell.
Smith, Susan J. 1989) *The politics of "race" and residence: Citizenship, segregation, and white supremacy in Britain*. Cambridge, Mass.: Polity.
Soja, Edward W. 1996. *Thirdspace: Journeys to Los Angeles and other real-and-imagined places*. Oxford, Eng.: Blackwell.
Steinberg, Stephen. 1981. *The ethnic myth: Race, ethnicity, and class in America*. New York: Atheneum.
Strauss, Anselm L. 1987. *Qualitative analysis for social scientists*. Cambridge, N.Y.: Cambridge University Press.
Thompson, John B. 1984. *Studies in the theory of ideology*. Cambridge, Mass.: Polity.
Turner, Susan C. 1997. Barriers to a better break: Employers discrimination and spatial mismatch in metropolitan Detroit. *Journal of Urban Affairs* 12 (2): 123-41.

Turner, Margery Austin, and Ellen Ingrid Gould. 1997. *Location, location, location: How does neighborhood environment affect the well-being of families and children?* Discussion paper. Washington, D.C.: Urban Institute.
U.S. Bureau of the Census. 1995. *Census of Population and Housing, 1990 [United States]: Summary Tape File 4A.* Washington D.C.: U.S. Government Printing Office.
U.S. Bureau of the Census. 1997. *U.S. Census Data, Database: C90STF3A* [http://www.census.gov/cdrom/lookup].
U.S. Department of Housing and Urban Development. 2001. [http://www.hud.gov/].
Valentine, Gill. 1992. Images of danger: Women's sources of information about the spatial distribution of male violence. *Area* 21 (1): 22-29.
Valentine, Gill, Tracey Skelton, and Deborah Chambers. 1998. Cool places: An introduction to youth and youth cultures. In *Cool places: Geographies of youth cultures*, edited by Tracey Skelton and Gill Valentine, 1-32. London: Routledge.
Varady, David P., and Carole C. Walker. 2000. Vouchering out of distressed subsidized developments: Does moving lead to improvements in housing and neighborhood conditions? *Housing Policy Debate* 11 (1): 115-62.
Vergara, Camilo J. 1995. *The new American ghetto.* New Brunswick, N.J.: Rutgers University Press.
Waggoner, Dorothy. 1991. *Undereducation in America: The demography of high school dropouts.* New York: Auburn House.
Waldinger, Roger. 1997. Black/immigrant competition re-assessed: New evidence from Los Angeles. *Sociological Perspectives* 40 (3): 365-86.
Weber, Max. 1968. *Economy and society.* Vol. 1. New York: Bedminster Press.
Wial, Howard. 1991. Getting a good job: Mobility in a segmented labor market. *Industrial Relations* 30 (3): 396-416.
Williams, Raymond. 1958. *Culture and society: 1780-1950.* London: Chatto and Windus.
———. 1983. *Keywords: A vocabulary of culture and society.* Revised edition. New York: Oxford University Press.
———. 1989. *Resources of hope: Culture, democracy, socialism.* Edited by Robin Gable. London: Verso.
Wilkinson Frank, Ed. 1981. *The dynamics of labour market segmentation.* London: Academic Press.
Willis, Paul E. 1977. *Learning to labor: How working class kids get working class jobs.* Westmead, Eng.: Saxon House.
Wilson, David. 2001. Coloring the city: "Black-on-black violence" and the liberal discourse. *Tijdschift voor Economische en Sociale Geografie* 92 (3): 261-78.
Wilson, William Julius. 1987. *The truly disadvantaged: The inner city, the underclass, and public policy.* Chicago: University of Chicago Press.
———. 1996. *When work disappears: The world of the new urban poor.* New York: Vintage Books.

Witte, Larry. 1993. *A different American Dream: The low-income housing crisis in San Antonio*. Report. San Antonio, Tex.: Partnership for Hope.
Zeisel, John. 1981. *Inquiry by design: Tools for environmental-behavior research*. Monterey, Calif.: Brooks/Cole Publishing.
Zukin, Sharon. 1995. *The cultures of cities*. Oxford, Eng.: Blackwell.

Index

academic discourse, 14–22, 73, 83
acculturation, 33, 67, 83, 88
adolescence, 5
adult role models. *See* neighborhood effects
Alamo, 25
Alazan-Apache Housing Courts, 36, 45, 50

Bhabha, Homi K., 87
Bourdieu, Pierre, 3, 11

case study area selection, 95
census data, 91
child care, 19, 61, 65
Chinatown (Vancouver), 12
churches. *See* religious institutions
Cisneros, Henry, 28
clothing, 80
Community Development Block Grants, 16
cultural assimilation. *See* acculturation
cultural capital, 3, 11, 46, 86
cultural codes, 3, 78, 79
cultural colonization, 70
cultural exclusion, 19, 86
cultural hybridity, 87
cultural identity, 3, 4, 10, 19
cultural imperialism, 88
cultural representation, 2, 3, 10
culture, 2, 4, 10, 15, 70; definition of, 2
culture of poverty, 4, 18, 61, 85

Daughters of the Republic of Texas, 25
day care. *See* child care
devaluation of labor, 53, 86–87
dialect, 46, 78
dysfunction, 4, 14, 60, 70, 73, 79

economics, 53, 70; neoclassical, 9
education, 12, 30, 56, 66, 81; school enrollment, 91
employment, 37–38, 93; nonemployment, 15, 30, 33, 91; temporary, 32, 51; underemployment, 26; unemployment, 26
enrollment. *See* education
ethnic typecasting, 44, 51, 77, 78–79
ethnographic studies, 18, 21, 22
exclusion: cultural, 5; labor market, 4, 9
external processes of cultural differentiation, 12

factor analysis. *See* principle component analysis
family, 52, 61, 63, 78–79, 81, 93
Free Trade Agreement of the Americas, 86

G8, 86
Galster, George, 17
Gautreaux Assisted Housing Program, 14, 16

gender roles, 10, 42, 43, 44, 45, 77, 78–79
Gibson-Graham, J.K., 85
globalization, 2, 86

harassment, 42
HOPE VI, 16
housing projects, 13, 36, 45, 50
housing subsidies, 14, 16, 89
human agency, 75

identity: cultural, 3, 4, 10; labor market, 3, 6, 10; transposing, 87; work, 44–45; working class, 11
identity formation: external representations, 41; internal processes, 41
ideology, 74; community-based, 76; external, 81; neighborhood-based, 74
imaginaries, 3, 87, 89
immigrants, 13
industrial restructuring, 9
intervention, 59–60, 66

jumping scale, 71, 74, 82–83

labor market, 3, 5, 9, 29, 86; demand side, 9, 85; performance expectations, 46; secondary, 83; segmentation, 11, 9–14, 44, 52, 85; supply side, 10, 12, 85
language. *See* dialect
Learning to Labor, 3
Lefebvre, Henri, 4
Lewis, Oscar, 4, 61

marginality, 4
media reports, 42
Mikelsons, Maris, 17
mixed-income housing, 16, 89
motherhood, 42, 64–65, 77, 81, 92
Moving to Opportunities, 16, 89

neighborhood effects, 14–22, 57, 70, 86; adult role models, 14, 15, 58, 62, 67–69; causality, 16–20, 32; peer group influences, 14, 15, 58; quantitative models, 21
neighborhood revitalization, 16
neighborhoods, 3, 4; cultural identity of, 19; employment characteristics, 32; ethnic characteristics of, 33; spatial confinement of, 42–43, 50, 67–69, 73, 78
neoclassical economics, 9
neoliberalism, 52, 70
networks: peer, 58; personal, 56
nonemployment. *See* employment

occupations, 45; health care, 45; primary, 49, 64, 86, 94; secondary, 44, 47, 48, 51, 53, 64, 86, 94; service industry, 11, 27
otherness, 10

parenting responsibilities, 42
pathology, 4, 15, 32, 83, 86
peer group influences. *See* neighborhood effects
Pinderhughes, Howard, 74
place discrimination, 13
police, 79
policy and planning initiatives, 16
political economy, 9
poverty, 93
principle component analysis, 91–92
Prospect Hill, 28
public discourse, 85–86

qualitative analysis, 96–100; institutional samples, 96; interviewing technique, 98; youth samples, 98 quantitative analysis, 91–96; principle component analysis, 91–92; regression analysis, 92–95

rap music, 87
regression analysis. *See* quantitative analysis
religious institutions, 56, 82
rescrypting, 86–87
Ruddick, Susan M., 5

San Antonio, 25–29; economy, 28; ethnic discrimination in, 27–29; population, 26; tourist industry, 25; unemployment, 26
San Antonio de Valero, 25
scale: urban, 77–78
scales of representation, 74
segmentation. *See* labor market
segmentation theory, 10, 52
segregation, 12, 28, 85, 89
self esteem, 47–48
service industry. *See* occupations
social hierarchy, 10
Soja, Edward W., 4
spatial confinement. *See* neighborhoods
spatial mismatch, 9, 10, 38

spatial scale, 70
Steinberg, Stephen, 4
structures of feeling, 11
style of dress, 80
suburbs, 14, 16, 89
symbols, 11, 86

teenage parenthood, 64, 77, 79, 92
Texas Workforce Commission, 79
transportation, 42
The Truly Disadvantaged, 1

U.S. Census, 29, 33, 91
U.S. Department of Housing and Urban Development, 16
underemployment. *See* employment
unemployment. *See* employment
upward mobility, 50
urban underclass, 4, 15, 18, 85

Vancouver's Chinatown, 12

Weber, Max, 10
welfare, 4, 50
Williams, Raymond, 2, 11
Willis, Paul E., 3
Wilson, William Julius, 1, 93
World Trade Organization, 86

About the Author

Harald Bauder is an assistant professor in the Geography Department at the University of Guelph. He has previously taught at Wayne State University in Detroit and completed postdoctoral research at the University of British Columbia in Vancouver. His research interests are in urban neighborhoods, issues of ethnic residential segregation, and labor market integration of immigrants and ethnic minorities. He has examined these issues in the United States, Canada, and Germany.